The American Invitational Mathematics Examination (AIME) Preparation

AIME PRACTICE TESTS

Lokman Gökçe, M.S., AIME - IMO Trainer

Andrew Carratu, Senior, USAMO Participant

Sinan Kanbir, Ph.D., Mathematics Education

December 2024

Copyright © 2024 MathTopia Academy LLC

All rights of this book and system are reserved. The text cannot be reproduced or published by electronic, mechanical, photocopy, or any recording system without the permission of MathTopia Press.

ISBN: 978-1-7356252-2-5 (Paper)

About the Authors

Lokman Gökçe, MS in Mathematics, is a distinguished educator with over two decades of experience in preparing middle and high school students for national and international math competitions, including AMC-8/10/12, AIME, and USA(J)MO. He provides remote education to students across the USA, Turkey, Azerbaijan, and the United Kingdom, adapting his methods to diverse learning needs and goals. Beyond teaching, Lokman shares his expertise as a YouTube content creator, offering insights on the International Mathematical Olympiad (IMO) and other math contests. His coaching has guided multiple IMO participants, fostering a new generation of mathematical talent worldwide.

Andrew Carratu, a Grade 12 student at Phillips Exeter Academy, is an accomplished young mathematician with an impressive competition record, including USAMO wins in 2024 (Grade 11) and 2023 (Grade 10), a USAJMO win in 2022 (Grade 9), and AIME qualifications dating back to 2021 (Grade 8). He also earned a silver medal at the Romanian Master of Mathematics in 2024 and consistently ranks on the top lists of the Harvard-MIT Mathematics Tournament (HMMT). Beyond competitions, Andrew creates video solutions for AMC and AIME problems and serves as a Teaching Assistant at MathTopia Academy, mentoring future mathematicians.

Sinan Kanbir, PhD., Mathematics Education, University of Wisconsin, specializing in mathematics education for future teachers at the elementary, middle, and high school levels. He has authored or co-authored eight books and numerous publications focused on enhancing math teaching and learning. He has been coaching AIME and USAMO students, helping them develop their mathematical talents. Additionally, he contributes to the Central Wisconsin Math League (CWML), MathCON, and the WMC State Math Contests.

Acknowledgments

We extend our heartfelt thanks to Mr.Bagbekov, MathTopia Academy instructor, for his invaluable feedback on every item, greatly enhancing this book's quality.

Special appreciation goes to James Gintner, founder of The Beauty of Math, for inspiring AMC 10–12 students to achieve AIME qualification with his unwavering support and encouragement.

We are grateful to MathPath Summer Camp students, particularly James Stewart, for verifying problems with enthusiasm, and to Suvid Bordi and Ayaan Garg for their energy and encouragement in shaping this book.

Lastly, we sincerely thank Dr. Karagoz for expertly resolving all LaTeX formatting issues, ensuring a professional and polished final product, and for teaching Andrew AIME topics and mentoring him in becoming a successful mathematician during his PhD studies.

Your contributions have been instrumental in making this book a reality. Thank you for your support and dedication.

To Future and Current AIME Mathletes

Welcome to a journey where persistence, creativity, and passion for problem-solving will define your path. Preparing for the American Invitational Mathematics Examination (AIME) is no small feat, but each step brings you closer to mastery, not just of math but of your own potential. Here are some essential tips to guide you on your way:

- Master the Fundamentals: Strong skills in algebra, geometry, trigonometry, and pre-calculus are the foundation for AIME success. Focus on understanding these core areas deeply so you can apply them creatively to complex problems.

- Practice with Purpose: Don't just solve problems—analyze them. Work through past AIME questions, noting recurring themes and techniques. Take time to understand each solution and reflect on why it works, so you can adapt similar strategies in new contexts.

- Cultivate Resilience: AIME questions are designed to challenge, sometimes beyond the first attempt. Embrace this as part of the process. Each struggle is a stepping stone toward growth, so stay persistent and keep pushing through even when it feels tough.

- Time Management is Key: AIME is as much about pacing as it is about problem-solving. Practice under timed conditions to develop a feel for how long you can spend on each question, balancing speed with accuracy.

- Learn from Mistakes: Treat every mistake as a valuable lesson. Carefully review errors to understand where you went wrong and how to avoid similar pitfalls. This reflection process will sharpen your intuition and problem-solving skills over time.

Remember, the AIME is more than just an exam—it's an opportunity to deepen your mathematical insight and push the boundaries of your problem-solving abilities. Whether you're aiming for the

USA Math Olympiad (USAMO) or simply looking to challenge yourself, the journey itself will be rich with rewards.

Best of luck, and may your hard work lead you to new heights!

My AIME Story

Andrew at Romanian Master in Mathematics (RRM) proudly received the silver medal as the USA team secured 1st place on March 1, 2024, in Bucharest, Romania.

My math story began long before 5th grade, but it was at my first notable national math contest—MATHCON in Chicago—that everything truly clicked. Competing at MATHCON was a transformative experience, filled with excitement and pride as I represented my school on a national stage. Traveling to Chicago, shining in the competition, and sharing the moment with friends made it unforgettable. One of the highlights of that event was meeting Dr. Kanbir, a problem writer for MATHCON. His passion for mathematics was contagious, leaving a lasting impression on me and inspiring me to aim higher and dream bigger.

In 7th grade, during the COVID-19 pandemic, I joined and successfully passed the AMC 10, marking the beginning of my AIME journey. A particularly memorable moment from that time was working with an early version of Dr. Kanbir's book *GeoTopia, Volume 1*. I vividly recall how one geometry problem he showed me before the exam ended up appearing on the test—and I solved it correctly! That small but significant moment helped pave the way for my AIME journey. This experience taught me the importance of taking every math and problem-solving opportunity seriously.

That spark of inspiration stayed with me when I later qualified for the AIME for the first time. At the time, I could barely solve one or two problems on past exams. Fortunately, my resourceful coach, Dr. Kanbir, introduced me to an internationally renowned AIME-IMO instructor, Mr. Lokman. Under his guidance, I worked diligently, attending weekly remote teaching sessions where he provided carefully curated, high-quality materials designed to push me to my best.

With Mr. Lokman's dedication, Dr. Kanbir's encouragement, and my mom's gentle but persistent support (for which I'm deeply grateful—without her, I wouldn't be here sharing my AIME story), I persevered. Through consistent practice and determination, I eventually achieved a score of 7.

Those early steps marked the beginning of an incredible journey that has taken me to remarkable milestones: becoming a two-time Math Olympiad Program (MOP) qualifier and competing in the Romanian Masters of Mathematics. MOP introduced me to an extraordinary community of mathematicians—a network of peers and mentors who have profoundly shaped and inspired me. I hope that one day, you too will have the chance to experience something similar.

This past year, I fell short of making the USA Math Team, but I've been making great progress and remain hopeful about achieving that goal in the future. As the saying goes, "The journey is more important than the destination." Each step along the way has been filled with valuable lessons, growth, and inspiration. No matter where you are on your mathematical journey, I encourage you to embrace the joy and excitement at every stage. Let your curiosity lead the way—you never know just how far it might take you!

Thank you for reading my AIME story. I hope you'll be inspired to write your own story soon!

Best,

Andrew Carratu

Senior at Phillips Exeter Academy

Contents

1 **American Invitational Mathematics Examination AIME**   1

    AIME Test 1 4

    AIME Test 2 11

    AIME Test 3 18

    AIME Test 4 25

    AIME Test 5 32

2 **Answers**   39

3 **AIME Practice Tests Solutions**   41

    Solutions of AIME Test 1 42

    Solutions of AIME Test 2 50

    Solutions of AIME Test 3 62

    Solutions of AIME Test 4 72

    Solutions of AIME Test 5 85

Chapter 1

American Invitational Mathematics Examination AIME

AIME Test 1

Instructions

1. This is a 15-question, 3-hour examination. All answers are integers ranging from 000 to 999, inclusive. Your score will be the number of correct answers; there is no partial credit or penalty for wrong answers.

2. Only scratch paper, graph paper, ruler, compass, and protractor are allowed. Calculators and computers are not permitted.

3. Figures in the test are not necessarily drawn to scale.

4. Be sure to record all of your answers and any other required information as instructed.

5. You will have 3 hours to complete this test.

AIME Test 1

1. x and y are integers that satisfy the equation

$$xy + 7x = 2x^2 + 3y + 1083.$$

What is the sum of all the possible values of x?

2. $ABCD$ is a convex tangential quadrilateral (it has an incircle tangent to every side). If $\angle A = \angle D = 90°$, $\angle B = 45°$, $AD = 12$, and the incircle touches BC at point E, then find $BE^2 + CE^2$.

3. a and b are positive integers such that $\log(a + 2b - 17) + \log(a + b - 5) = 2$. Find $a \cdot b$.

4. Linda and Arnold are playing a game. Linda will throw a pair of fair dice. Let S be the sum of the numbers on the top of the dice. If $S > 5$, Arnold will give Linda S dollars. If $S < 6$, Linda will give Arnold $2S$ dollars. The expected value of Linda's earnings is $\frac{m}{n}$ dollars, where m and n are relatively prime positive integers. Find $m + n$.

5. Find the number of sequences of ten numbers $a_1, a_2, ..., a_{10}$ such that each a_i is either 1 or 2, and there is some integer $1 \leq x \leq 7$ such that $a_x = 1, a_{x+1} = 2, a_{x+2} = 1$, and $a_{x+3} = 2$.

6. In the solid below, $ABCD$ and $EFGH$ are rectangles with $AB = 12$, $AD = 11$, $EF = 4$, $EH = 5$, and $AE = BF = CG = DH = 13$. If planes $ABCD$ and $EFGH$ are parallel, and lines AB and EF are also parallel, what is the volume of the solid?

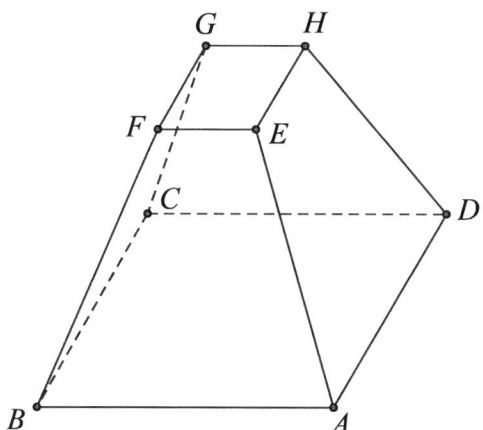

AIME Test 1

7. Let $a_1, a_2, a_3, \ldots$ be a positive integer sequence such that $a_1 = 1$ and

$$2a_{n+1} + 2a_n - 1 = (a_{n+1} - a_n)^2$$

for all positive integers n. Find the number of positive integers a_{999} can equal.

8. If the quartic equation $x^4 + 8x^3 + 15x^2 = 4x + n$ has 4 distinct real roots, what is the greatest integer value n can take?

9. Let $x, y,$ and z be positive real numbers such that

$$x + y + z = 21$$
$$\sqrt{x^2 + 16} + \sqrt{y^2 + 49} + \sqrt{z^2 + 81} = 29$$

If x can be expressed as $\dfrac{m}{n}$, where m and n are relatively prime positive integers, find $m + n$.

10. Let $ABCD$ be a square and P be the point inside the square such that $PB = 10$, $PC = 8$, and $PD = 8$. The area of $ABCD$ can be written as $m + n\sqrt{p}$, where m, n, and p are positive integers, and p is not divisible by the square of any prime. Find $m + n + p$.

11. Let $\triangle ABC$ have side lengths $AB = 20$, $BC = 9$, and $CA = 15$ and P be an arbitrary point on side $\overline{BC}$. Let ω_1 and ω_2 be the inscribed circles of $\triangle ABP$ and $\triangle ACP$, respectively. The common exterior tangent of ω_1 and ω_2 (other than BC) intersects with AP at Q. Find AQ.

12. Hamilton is standing on vertex A of regular hexagon $ABCDEF$. Every minute, he will walk to one of the two vertices adjacent to the one he is standing on. When he reaches point D, he will stop. In how many different ways can he reach point D after exactly 13 steps?

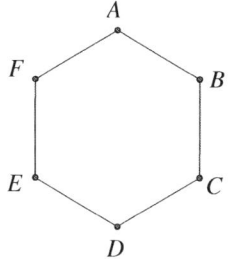

AIME Test 1

13. A positive integer greater than 1 is called a *good number* if it can be divided by any positive integer less than its square root. How many good numbers are there?

14. Find the number of integers n such that $0 \leq n < 2021$ and 2021 divides $n^6 - 1$.

15. In $\triangle ABC$, $\angle BAC = 120°$ and $AB : AC = 7 : 8$. Let P be an arbitrary point in $\triangle ABC$ on the angle bisector of $\angle BAC$. The minimum possible value of the ratio $\frac{BP}{CP}$ can be written as $\frac{\sqrt{m}}{n}$, where m and n are relatively prime positive integers. Find $m + n$.

AIME Test 1

Evaluation Sheet for Test 1

Question	Your Answer	Correct Answer	Check ☑ ☒ ☐	Notes
1				
2				
3				
4				
5				
6				
7				
8				
9				
10				
11				
12				
13				
14				
15				
Total Points				

AIME Test 2

Instructions

1. This is a 15-question, 3-hour examination. All answers are integers ranging from 000 to 999, inclusive. Your score will be the number of correct answers; there is no partial credit or penalty for wrong answers.

2. Only scratch paper, graph paper, ruler, compass, and protractor are allowed. Calculators and computers are not permitted.

3. Figures in the test are not necessarily drawn to scale.

4. Be sure to record all of your answers and any other required information as instructed.

5. You will have 3 hours to complete this test.

AIME Test 2

1. Real numbers $x, y,$ and z satisfy the following inequality system:

$$\begin{aligned} 1 &\leq 3x + y - z \leq 11 \\ -3 &\leq x - 3y + 5z \leq 7 \\ -4 &\leq x - 2y + z \leq 6 \end{aligned}$$

If $x + y + z$ is an integer, find the sum of all possible values of $x + y + z$.

2. In $\triangle ABC$, $AB = 187$ and $BC - CA = 119$. The length of the altitude from C to AB is 204, and $\cot \angle C = \dfrac{m}{n}$, where m and n are relatively prime positive integers. Find $m + n$.

3. When people buy tickets at a movie theater, their birthdays are recorded. A one-time free ticket is given to the first person in line who was born on the same day as some customer who bought a ticket before. Walking up to the ticket office, Andrew can enter the line wherever he wants. For example, he may choose to be the 1st person in line or the 1000th, or somewhere in between. Andrew does not know the birthday of anyone else in the line. In order to maximize his chances of getting the free ticket, Andrew queues as the nth person in the line. Find n.

(Assume the odds of a person having a birthday on any of the 365 days in a year to be equal, and that no one has a birthday on leap day.)

AIME Test 2

4. Let $ABCD$ be a tetrahedron with $AB = AC = 3\sqrt{13}$, $BC = 18$, $AD = 10$, and $BD = CD = 11$. Find the volume of the tetrahedron.

5. If $m = \sum_{n=1}^{300} \left\lfloor \dfrac{2^n}{5} \right\rfloor$, what is $2^{301} - 5m$?

($\lfloor x \rfloor$ equals the largest integer not greater than x.)

6. The solutions to the equation $z^{24} + 64z^{12} + 4096 = 0$ are the vertices of a convex polygon in the complex plane. The area of the polygon is $(\sqrt{a} - \sqrt{b}) \cdot \cos(5°)$, where a and b are positive integers. Find $a + b$.

7. Positive integers $a, b,$ and c satisfy the equation

$$ab(b^3 - a^3) + bc(c^3 - b^3) + ca(a^3 - c^3) + a^2b^2(b-a) + b^2c^2(c-b) + c^2a^2(a-c) = 450.$$

Find abc.

8. Let $S = \binom{40}{0} + \binom{40}{4} + \binom{40}{8} + \cdots + \binom{40}{40}$. Find the remainder when S is divided by 1000.

9. a, b, c positive numbers and $a + b + c = 6$. Minimum value of

$$\frac{\log_a b}{2a - 3b + 20} + \frac{\log_b c}{b - c + 18} + \frac{\log_c a}{28 - c - 4a}$$

is $\dfrac{m}{n}$, where m and n are relatively coprime positive integers. Find $m + n$.

AIME Test 2

10. What is the remainder of the product of n integers satisfying the conditions $1 \leq n \leq 1000$ and $\gcd(n, 1000) = 1$, divided by 1000?

11. A, B, C, D, and E lie on a circle in that order. $\angle ABE = \angle DBC$ and $\angle BAC = \angle DAE$. BC and AE intersect at F. $AB = 6$, $CE = 5$, and the circumradius of triangle ABF is $\frac{m}{n}$, where m and n are relatively prime positive integers. Find $m + n$.

12. The polynomial $P(x) = x^3 + (2n - 5)x^2 + (n^2 + 4n + 3)x + n + 3$ has three positive real roots given as x_1, x_2, x_3. If $\arctan(x_1) + \arctan(x_2) + \arctan(x_3) = \dfrac{3\pi}{4}$, what is the absolute value of the product of the coefficients of $P(x)$?

13. If the n-th root of a positive integer with n digits in the decimal number system is less than n from the sum of the digits of this number, the number is called *wonderful number* where $n \geq 2$. For example, $\sqrt[3]{216} = 2+1+6-3$ and 216 is a wonderful number. What is the remainder of the sum of the wonderful numbers less than 10000 divided by 1000?

14. A 16-person chess tournament will be held, with two students from 8 schools each participating. 8 simultaneous chess matches can be arranged in N different ways so that students from the same school do not match with each other. What is the remainder after dividing N by 1000?

15. $\triangle ABC$ has the length of sides $AB = 5$, $BC = 8$, $CA = 7$ and a point P is taken on the plane of $\triangle ABC$. Let $PA = x$, $PB = y$, $PC = z$. What is the minimum integer value of $5x + 7y + 8z$?

AIME Test 2

Evaluation Sheet for Test 2

Question	Your Answer	Correct Answer	Check ☑ ☒ ☐	Notes
1				
2				
3				
4				
5				
6				
7				
8				
9				
10				
11				
12				
13				
14				
15				
Total Points				

AIME Test 3

Instructions

1. This is a 15-question, 3-hour examination. All answers are integers ranging from 000 to 999, inclusive. Your score will be the number of correct answers; there is no partial credit or penalty for wrong answers.

2. Only scratch paper, graph paper, ruler, compass, and protractor are allowed. Calculators and computers are not permitted.

3. Figures in the test are not necessarily drawn to scale.

4. Be sure to record all of your answers and any other required information as instructed.

5. You will have 3 hours to complete this test.

AIME Test 3

1. The table below shows how many hours it took Geralt of Rivia, Vesemir, Cirilla and Yennefer to complete a task as two people each.

	Cirilla	Yennefer
Geralt of Rivia	m/n	15
Vesemir	8	10

   Geralt of Rivia and Cirilla can complete the same task in m/n hours, whose m, n are co-prime positive integers. What is $2m + n$?

2. Points $A(-2, 5)$, $B(2, 1)$, $C(8, 1)$ are given on the analytical plane. If $O(a, b)$, $H(c, d)$, $G(e, f)$ are circumcenter, orthocenter, centroid of the triangle ABC, respectively, what is $a^2 + b^2 + c \cdot d + e + f$?

3. Let α, β, γ be roots of the equation $x^3 - 6x^2 - 331x - 8 = 0$. What is $\sqrt[3]{\alpha} + \sqrt[3]{\beta} + \sqrt[3]{\gamma}$?

4. a, b, c are positive integers satisfy the following equation

$$\frac{1}{a} + \frac{1}{b} + \frac{1}{c} = \frac{7}{24}.$$

What is the maximum value of $a + b + c$?

5. $ABCD$ is a square and it's divided into 9 equal squares. When going from corner A to corner C, only east, north or northeast directions can be used. At each point, the direction to go is chosen randomly with equal probability. For example, when at point B, the probability of going north is 1, since it is only possible to move in the north direction. The probability of going through point X on the way from A to C is m/n, whose m and n are co-prime positive integers. What is $m + n$?

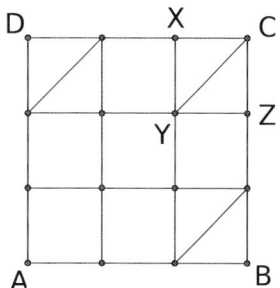

6. $ABCDEF$ is a regular hexagon and P is a point inside of the hexagon. The areas of triangles APF, BPA, CPB are 27, 24, 15, respectively. What is the area of the hexagon?

AIME Test 3

7. The Paradis Island scout unit consists of 1700 people. 91% of the scouts like Rogue Titan, 79% like Armored Titan, 77% like Colossal Titan, 86% like Female Titan. If the minimum and the maximum number of people in the scout unit that likes all four titans are a and b, respectively, then what is $b - a$?

8. The f_n Fibonacci sequence is defined as $f_1 = f_2 = 1$ and for $n \geq 1$, $f_{n+2} = f_{n+1} + f_n$.

$$\sum_{n=1}^{14} \left(\frac{2n}{f_n} - \frac{n \cdot f_{n+2}}{f_{n+1} \cdot f_n} - \frac{1}{f_{n+1}} \right) = \frac{m}{n},$$

where m and n are co-prime positive integers. What is $m + n$?

9. In the decimal system, if the rightmost digit of a positive integer is replaced by the leftmost digit, the resulting number is equal to 3 times the first number. What is the remainder of the division by 1000 of the smallest number satisfying this condition?

10. The side lengths of triangle ABC are integers and $\angle BAC = \alpha$, $\angle ABC = \beta$, $\angle BCA = \gamma$. If $\sin\beta = 4\sin\alpha\cos\gamma$, what is the minimum value of the perimeter of triangle ABC?

11. A regular pentagonal pyramid is a solid with a regular pentagonal base and equilateral triangular lateral surfaces. The volume of a regular pentagonal pyramid with a side length of 6 is $m + n\sqrt{p}$. Here m, n, p are positive integers and p is a square-free number. What is $m + n + p$?

12. A function f defined on real numbers satisfy the equation
$$f(x+y) + f(x-y) = 2f(x) + 2f(y) - 6y$$
for every real number x, y. If $f(1) = 4$, what is $f(-24)$?

AIME Test 3

13. There are 1000 cities in a country. There are 250000 roads in total between cities. If there are direct roads from one city to all other cities, we call that city *metropolis*. What is the maximum number of metropolises?

14. P is an arbitrary point on the circumcircle of the equilateral triangle ABC. If radius of the circumcircle is $\sqrt{6}$, what is $PA^4 + PB^4 + PC^4$?

15. p and $p^4 - 35p^3 + 365p^2 - 1225p + 1259$ are prime numbers. What is sum of all possible values of such p primes?

AIME Test 3

Evaluation Sheet for Test 3

Question	Your Answer	Correct Answer	Check ☑ ☒ ☐	Notes
1				
2				
3				
4				
5				
6				
7				
8				
9				
10				
11				
12				
13				
14				
15				
Total Points				

AIME Test 4

Instructions

1. This is a 15-question, 3-hour examination. All answers are integers ranging from 000 to 999, inclusive. Your score will be the number of correct answers; there is no partial credit or penalty for wrong answers.

2. Only scratch paper, graph paper, ruler, compass, and protractor are allowed. Calculators and computers are not permitted.

3. Figures in the test are not necessarily drawn to scale.

4. Be sure to record all of your answers and any other required information as instructed.

5. You will have 3 hours to complete this test.

AIME Test 4

1. Let $s(n)$ denote the sum of the digits of n, when written in base 10. For example, $s(123) = 6$ and $s(2024) = 8$. Let X be the number of positive integers n less than 10^5 that have the property that $s(n) > 10$. Find the first 3 digits of X.

2. From point P on the circumcirle of $\triangle ABC$, perpendiculars $\overline{PX}, \overline{PY}$ are drawn to lines AC, BC, respectively. P is in the part of arc BC that does not contain point A. If $PX = 45$, $PB = 28$, $PY = 20$, then what is length of PA?

3. What is the sum of the 5-th powers of all the real roots of the equation

$$\log_x(x^3 - 2x^2 - 5x + 14) = 2 \ ?$$

AIME Test 4

4. a, b are co-prime positive integers. What is the sum of all the different values

$$\gcd(a+b, a^2 + 17ab + b^2)$$

can take?

5. One of the (x, y) integer pairs satisfying the inequality $|x - y| + |x + y| \leq 20$ is chosen randomly and with equal probability. The probability that $x^2 + y^2 > 35$ is m/n, which m, n co-prime positive integers. What is $m + n$?

6. $ABCD$ is a rectangle and $AD = 10$, $AB = 12$. Points E, F, G, H on the sides AB, BC, CD, DA, respectively such that $AE = DH = 6$, $EH \parallel FG$. The maximum value of the area of the quadrilateral $EFGH$ is m/n, which m, n co-prime positive integers. What is $m + n$?

7. In isosceles trapezoid $ABCD$, $AB = 2024$, $CD = 1$ and $AB \parallel CD$. If AC, BC lengths are integer. What is the sum of the different integer values that AC can take?

8. We are given n metal weights with positive integer masses (in grams). A brick weighing $k > 0$ grams is glued to the left side of a double-pan balance scale. Our goal is to add some weights to either side of the scale to make it balanced. For example, if we have two weights of 1 gram and 5 grams, and $k = 4$, we can balance the scale by placing the 1 gram weight on the left side and the 5 gram weight on the right side. As it turns out, our n metal weights make it possible to do this for any value of k between 1 and 4000 inclusive. What is the minimum possible value of n?

9. The tetrahedral sequence of numbers is formed by the number of points: In the first step, there is 1 point. In the $n+1$-th step, a regular tetrahedron with one edge of n units is drawn. Newly added points are on vertices, edges, and surfaces. Among these points, the closest ones are 1 unit apart. The first four tetrahedral numbers are 1, 4, 10, 20, which are shown in the figure below. What term is the number 1873200 in the tetrahedral sequence of numbers?

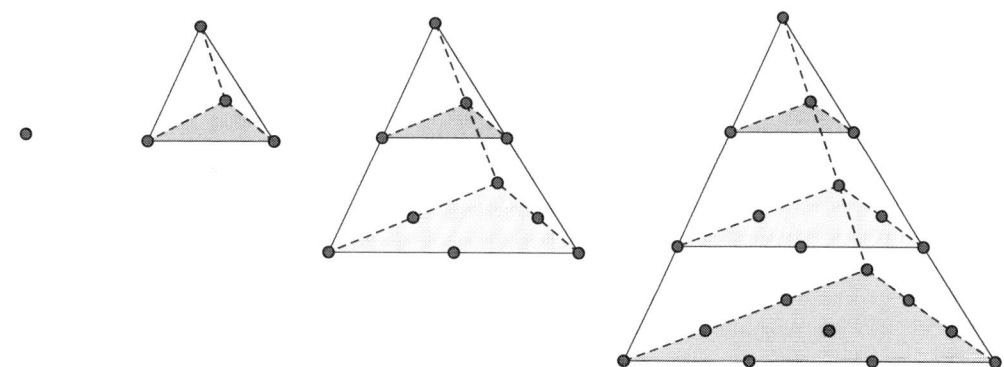

10. Y is the midpoint of segment XZ. We draw parallel lines x, y, and z through X, Y, and Z respectively so that all 3 lines are perpendicular to XZ. Circle O intersects with x at points A and B, intersects with y at points C and F, and intersects with z at points D and E so that $ABCDEF$ is a convex hexagon. $AB = 2$, $CF = 5$, $DE = 1$, and $XY = \frac{a\sqrt{b}}{c}$, where a, b, and c are positive integers with a and c relatively prime and b is not divisible by the square of a prime. Find $a + b + c$.

11. How many ways are there to color the first 7 positive integers so that each one is colored either red or blue, and there are no 3 integers $1 \le a < b < c \le 7$ all the same color forming an arithmetic progression?

12. Positive real numbers a, b, c, and d satisfy

$$a + b + c + d = 7$$
$$ab + bc + cd + da = 12$$
$$a^2 + 2b^2 + 3c^2 + 4d^2 = k$$

and the minimum possible value of k is $\frac{m}{n}$ where m and n are relatively prime positive integers. Find $m + n$.

13. We have $n+1$ different colors of paint. The faces of a regular n-gon pyramid will be painted using these colors. The lateral faces of the pyramid are isosceles triangles. The colorings obtained from one another as a result of the rotation of the object are considered identical. The faces of the pyramid can be painted in $S(n)$ different ways, with each face being a different color.
$$\frac{1}{s(3)} + \frac{1}{S(4)} + \frac{1}{S(5)} + \cdots + \frac{1}{S(2023)} + \frac{1}{2024!} = \frac{m}{n},$$
where m and n relatively co-prime positive integers. What is $m+n$?

14. In triangle ABC with $|BC| = 6$, squares $BADE$, $CBFG$, and $ACHK$ are constructed on the sides of the triangle and extending outward. The points D, E, F, G, H, K are concyclic. Let S be the sum of all distinct possible perimeters of triangle ABC. What is the integer part of S?

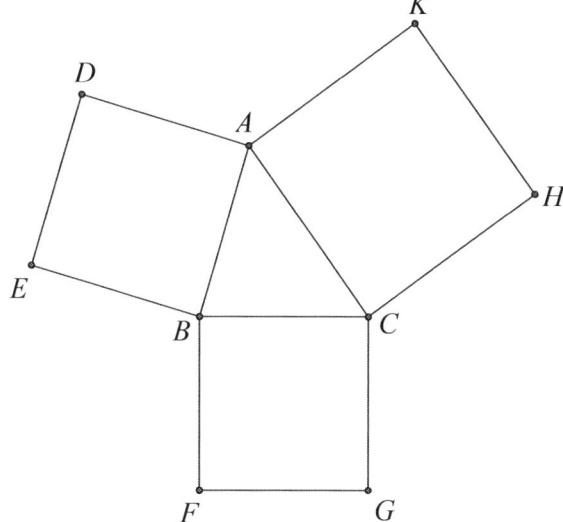

15. Let $n \geq 0$ be an integer and let $f(n)$ be the remainder when the smallest prime divisor of $6^{2^n} + 1$ is divided by 32. Find the value of
$$\sum_{n=0}^{100} f(n).$$

AIME Test 4

Evaluation Sheet for Test 4

Question	Your Answer	Correct Answer	Check ☑ ☒ ☐	Notes
1				
2				
3				
4				
5				
6				
7				
8				
9				
10				
11				
12				
13				
14				
15				
Total Points				

AIME Test 5

Instructions

1. This is a 15-question, 3-hour examination. All answers are integers ranging from 000 to 999, inclusive. Your score will be the number of correct answers; there is no partial credit or penalty for wrong answers.

2. Only scratch paper, graph paper, ruler, compass, and protractor are allowed. Calculators and computers are not permitted.

3. Figures in the test are not necessarily drawn to scale.

4. Be sure to record all of your answers and any other required information as instructed.

5. You will have 3 hours to complete this test.

AIME Test 5

1. Two of the positive integers from 1 to $2n+1$ (including 1 and $2n+1$) are chosen. The selected numbers can be the same. If the probability is $\dfrac{320}{441}$ of the product of the two chosen numbers is an even number, what is n?

2. In the semicircle, $\overline{AB}$ is a diameter with $AB = 8$. Points C and D are on the semicircle such that $BC = CD = 2$ and point C is adjacent to point B. What is $\overline{AD}^2$?

3. A unit cube is filled with $v < \frac{1}{2}$ cubic units of water. The cube is tilted so that the surface of the water forms an equiangular hexagon $ABCDEF$, such that $AB = CD = EF = 2BC = 2DE = 2FA$. Then, $v = \frac{m}{n}$, where m and n are relatively prime positive integers. Find $m+n$.

4. Bob makes a bag of red and blue balls as follows:

   1. Start with an empty bag.

   2. Flip a fair coin to decide between red and blue, then add a ball of the color to the bag.

   3. If there are 3 balls in the bag, end the process immediately. Otherwise, flip a fair coin to decide between going back to step 2, or ending this process.

   Bob then hands this bag to Alice. She reaches into the bag and pulls out a ball. It is colored red. She places the ball back into the bag. She then reaches into the bag and pulls out another ball. The probability of this ball being red as well is $\frac{m}{n}$, where m and n are positive integers. Find $m + n$.

5. A 6×5 table containing the natural numbers from 1 to 30 has been provided. One number is selected from each column and these five numbers are multiplied together. For example, $1 \cdot 2 \cdot 8 \cdot 9 \cdot 30 = 4320$. The sum of all these products is denoted as S. What is the remainder when S is divided by 1000?

1	2	3	4	5
6	7	8	9	10
11	12	13	14	15
16	17	18	19	20
21	22	23	24	25
26	27	28	29	30

6. Let ABC be a triangle, and let D be the midpoint of $\overline{BC}$. The incircle ω of $\triangle ABC$ intersects $\overline{AD}$ at two distinct points, which trisect the median $\overline{AD}$. Circle ω is tangent to $\overline{BC}$ at point E. Given that $ED = 8$, and D is between points E and C. If the area of ω is $\frac{a\pi}{b}$, where a and b relatively co-prime positive integers, what is $a + b$?

AIME Test 5

7. Given that $(x^{2022} + x^3 - 1)^{2023} = a_0 + a_1 x + a_2 x^2 + a_3 x^3 + \cdots + a_{2022 \cdot 2023} x^{2022 \cdot 2023}$. Find the sum

$$S = a_0 + a_3 + a_6 + \cdots + a_{2022 \cdot 2023}.$$

8. Given that m and n are positive real numbers, there is no real number pair (x, y) that satisfies the system of equations:

$$x^3 + y^3 = 16$$
$$mx + y = n$$

What is the smallest positive integer value that $m^2 + n^2$ can take?

9. A student plays a game as follows. They first choose a real number p between 0 and 1, inclusive. They then receive a weighted coin that lands heads with probability p and tails with probability $1 - p$. They flip it 7 times, and they win if it comes up heads exactly 3 times. The value of p the student should choose to maximize their odds of winning is $\frac{m}{n}$, where m and n are relatively prime positive integers. Find $m + n$.

10. We have that a and b are positive real numbers such that $a + b = 2408$ and
$$\log_a(\log_b(a)) = \log_b(\log_a(\sqrt[1024]{b}))$$
Find $b - 300a$.

11. Let $ABCD$ be a tetrahedron, and M be the midpoint of AB. We have $CD = 14$, $CA = 13$, $AD = 15$, $CM = 5$, and $MD = \sqrt{109}$. Find the largest possible integer value that the perimeter of triangle CBD can be.

12. Let the Fibonacci numbers be defined $F_1 = 1$, $F_2 = 1$, and $F_{n+2} = F_{n+1} + F_n$ for all $n \geq 1$. If
$$\sum_{n=5}^{\infty} \frac{1 + (-1)^n F_n F_{n+2}}{F_n F_{n+1} F_{n+2}} = \frac{m}{n}$$
Where m and n are relatively prime integers, find $-mn$.

AIME Test 5

13. In Triangle ABC with $BC = 5$, $CA = 27$, and $\angle ABC = 90°$, let E be a point on segment CA. Let ω be the circle passing through B and tangent to CA at E. ω intersects segment BC at a point D not equal to B, and we have that $BD = DE = \frac{m}{n}$, where m and n are relatively prime positive integers. Find $m + n$.

14. Non-constant polynomials $P(x)$, $Q(x)$, and $R(x)$ with integer coefficients satisfy the following conditions:

    - All 3 polynomials have the same degree.
    - For all real numbers x, $(P(x))^2 + (Q(x))^2 = (R(x))^2$.
    - For all real numbers x, $P(Q(x)) - P(0) + Q(P(x)) - Q(0) + 4x = R(R(x)) - R(0) + 6$.

    Find the sum of the three smallest values of $|R(0)|$.

15. Starting from vertex A of a regular n-gon with a side length of 1 unit, moving 1 unit along the edges in the positive direction leads to point B, moving 2 units leads to point C, moving 4 units leads to point E, and moving 5 units leads to point F. (It is possible to return to points A, B, or C as a result of the movement.) Given that the areas of triangles ABF and ACE are equal, what is the sum of the possible values of n?

Evaluation Sheet for Test 5

Question	Your Answer	Correct Answer	Check ✓ ☒ ☐	Notes
1				
2				
3				
4				
5				
6				
7				
8				
9				
10				
11				
12				
13				
14				
15				
Total Points				

AIME Test 5

Answers

Test 1

1. 192
2. 216
3. 036
4. 014
5. 357
6. 816
7. 500
8. 019
9. 026
10. 127
11. 013
12. 486
13. 007
14. 012
15. 005

Test 2

1. 063
2. 089
3. 020
4. 108
5. 752
6. 288
7. 120
8. 232
9. 007
10. 001
11. 419
12. 765
13. 773
14. 672
15. 080

Test 3

1. 251
2. 112
3. 000
4. 629
5. 103
6. 108
7. 748
8. 241
9. 857
10. 024
11. 059
12. 504
13. 292
14. 648
15. 040

Answers

Test 4

1. 970
2. 063
3. 275
4. 024
5. 773
6. 247
7. 864
8. 009
9. 223
10. 017
11. 016
12. 025
13. 007
14. 043
15. 143

Test 5

1. 010
2. 049
3. 223
4. 089
5. 545
6. 121
7. 001
8. 018
9. 010
10. 301
11. 053
12. 120
13. 021
14. 105
15. 015

Chapter 3

AIME Practice Tests Solutions

Solutions of AIME Test 1

1. x and y are integers that satisfy the equation

$$xy + 7x = 2x^2 + 3y + 1083.$$

What is the sum of all the possible values of x?

Solution: The problem is with the approach of leaving one of the variables alone. We can observe that it is easier to leave the variable y alone. $xy - 3y = 2x^2 - 7x + 1083$ and $y(x-3) = 2x^2 - 7x + 1083$. So $y = \dfrac{2x^2 - 7x + 1083}{x-3}$. From polynomial division, we find $2x^2 - 7x + 1083 = (2x-1)(x-3) + 1080$. Thus,

$$y = (2x-1) + \frac{1080}{x-3}$$

and we conclude that $(x-3) \mid 1080$. With prime factorization, we write $1080 = 2^3 \cdot 3^3 \cdot 5$. From the formula for the number of integer divisors, there are $2(3+1)(3+1)(1+1) = 64$ different $x - 3$ values that they sum is 0. Like $x - 3 = 1$, $x - 3 = -1, \ldots$, $x - 3 = 1080$, $x - 3 = -1080$. Therefore, the sum of all x values is $64 \cdot 3 = 192$.

2. $ABCD$ is a convex tangential quadrilateral (it has an incircle tangent to every side). If $\angle A = \angle D = 90°$, $\angle B = 45°$, $AD = 12$, and the incircle touches BC at point E, then find $BE^2 + CE^2$.

Solution: Define F, G, and H to be the incircle tangency points to CD, DA, and AB respectively, and let O be the center of the incircle. We can see that $OFDG$ and $OGAH$ are squares, thus $FD = DG = GA = AH = 6$, and if we draw I on AB such that CI and AB are perpendicular we see that CIB is an isosceles right triangle with $CI = 12$ and thus $CB = 12\sqrt{2}$. Now, let $CF = x$. Then, CE and IH both equal x as well, and since $BI = 12$ we obtain that $EB = HB = 12 + x$. But we have that $CE + EB = CB$, and plugging in gives us $x + (12 + x) = 12\sqrt{2}$, which simplifies to $x = 6\sqrt{2} - 6$. So $CE = 6\sqrt{2} - 6$ and

42

$BE = 6\sqrt{2} + 6$, and we can now compute $BE^2 + CE^2 = 216$.

3. a and b are positive integers such that $\log(a + 2b - 17) + \log(a + b - 5) = 2$. Find $a \cdot b$.

Solution: First, we use a logarithm rule: $\log((a + 2b - 17)(a + b - 5)) = 2$. This clearly implies that $(a + 2b - 17)(a + b - 5) = 100$. Now we can just go through all the different factors of 100 until we find one that gives a positive integer solution (a, b). As it turns out, the only solution is when $a + 2b - 17 = a + b - 5 = 10$, yielding the solution $(a, b) = (3, 12)$, meaning $ab = 36$.

4. Linda and Arnold are playing a game. Linda will throw a pair of fair dice. Let S be the sum of the numbers on the top of the dice. If $S > 5$, Arnold will give Linda S dollars. If $S < 6$, Linda will give Arnold $2S$ dollars. The expected value of Linda's earnings is $\frac{m}{n}$ dollars, where m and n are relatively prime positive integers. Find $m + n$.

Solution: Clearly there are 36 different possibilities for what the dice can show, so to obtain the expected value of what Linda will earn we can just add up the earnings over all 36 cases and then divide by 36. There is 1 way to roll a 2, 2 ways to roll a 3, 3 ways to roll a 4, ..., up to 2 ways to roll an 11 and 1 way to roll a 12. Thus, Linda's total earnings are
$1 \cdot (-4) + 2 \cdot (-6) + 3 \cdot (-8) + 4 \cdot (-10) + 5 \cdot 6 + 6 \cdot 7 + 5 \cdot 8 + 4 \cdot 9 + 3 \cdot 10 + 2 \cdot 11 + 1 \cdot 12 = 132$,
thus the expected value she will earn is just $\frac{132}{36} = \frac{11}{3}$, giving an answer of 14.

5. Find the number of sequences of ten numbers $a_1, a_2, ..., a_{10}$ such that each a_i is either 1 or 2, and there is some integer $1 \leq x \leq 7$ such that $a_x = 1, a_{x+1} = 2, a_{x+2} = 1$, and $a_{x+3} = 2$.

Solutions of AIME Test 1

Solution: So we want to count the number of ten-number sequences that include the consecutive subsequence $1, 2, 1, 2$. It is easier to complementary count the number of ten-number sequences that don't include that subsequence. Let the number of n-number sequences not including "1,2,1,2" be a_n. Then our answer must be $2^{10} - a_{10}$. Clearly $a_1 = 2$, $a_2 = 4$, $a_3 = 8$, and $a_4 = 15$. Now, we will come up with a recursive formula for a_n. If the sequence starts with a 2, there are just a_{n-1} ways to fill in the rest of the sequence. If it starts with $1, 1$, this is clearly the same as the number of ways to write an $n - 1$ number sequence without the consecutive subsequence $1, 2, 1, 2$, assuming we start with the number 1, which as we have just shown equals $a_{n-1} - a_{n-2}$. If it starts with $1, 2, 2$, the number of ways to finish is just a_{n-3}. Finally, if it starts with $1, 2, 1, 1$, this is just the number of ways to write an $n - 3$ number sequence without the consecutive subsequence $1, 2, 1, 2$, assuming we start with the number 1, which equals $a_{n-3} - a_{n-4}$. Adding this all up, we obtain that $a_n = 2a_{n-1} - a_{n-2} + 2a_{n-3} - a_{n-4}$. Now, we can just extend the first 4 terms up until we reach the 10th term: $a_5 = 28$, $a_6 = 53$, $a_7 = 100$, $a_8 = 188$, $a_9 = 354$, and $a_{10} = 667$, giving a final answer of $1024 - 667 = 357$.

6. In the solid below, $ABCD$ and $EFGH$ are rectangles with $AB = 12$, $AD = 11$, $EF = 4$, $EH = 5$, and $AE = BF = CG = DH = 13$. If planes $ABCD$ and $EFGH$ are parallel, and lines AB and EF are also parallel, what is the volume of the solid?

Solution: Drop altitude from E to $ABCD$, call the foot of this altitude I. Then, drop the altitudes from I to AB and AD, call the feet of these altitudes J and K respectively. We have that $AJ = \frac{12-4}{2} = 4$ and $AK = \frac{11-5}{2} = 3$, thus $AI = 5$, thus $EI = 12$. Now, we can separate the solid into 9 solids by drawing the planes perpendicular to $ABCD$ and passing through EF, FG, GH, and HE respectively. This forms 4 rectangle pyramids with base area 12 and height 12, 2 triangular prisms with base area 18 and height 4, 2 triangular prisms with base area 24 and height 5, and 1 rectangular prism with base area 20 and height

12. Thus, the total volume is

$$4 \cdot \frac{144}{3} + 2 \cdot 18 \cdot 4 + 2 \cdot 24 \cdot 5 + 20 \cdot 12 = 816.$$

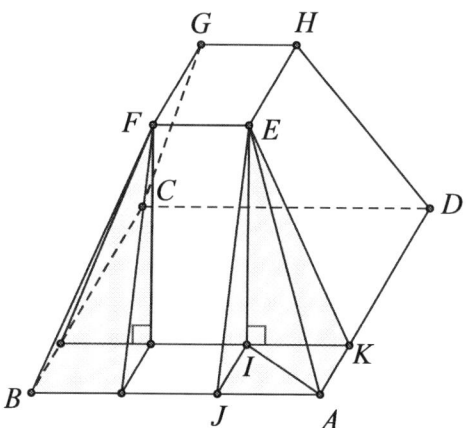

7. Let $a_1, a_2, a_3, \ldots$ be a positive integer sequence such that $a_1 = 1$ and

$$2a_{n+1} + 2a_n - 1 = (a_{n+1} - a_n)^2$$

for all positive integers n. Find the number of positive integers a_{999} can equal.

Solution: Through experimenting with what values a_2, a_3, etc. can be, we guess that if $a_n = k^2$ for some k, then a_{n+1} is either $(k+1)^2$ or $(k-1)^2$. Indeed, this is correct, as plugging in two consecutive squares into the equation causes everything to cancel:

$$((k+1)^2 - k^2)^2 - 2(k+1)^2 - 2k^2 + 1 = 4k^2 + 4k + 1 - 4k^2 - 4k - 2 + 1 = 0$$

As a result, it is easy to see that a_{999} must be one of the numbers $1^2, 3^2, 5^2, 7^2, \ldots 999^2$ meaning 500 different values.

8. If the quartic equation $x^4 + 8x^3 + 15x^2 = 4x + n$ has 4 distinct real roots, what is the greatest integer value n can take?

Solutions of AIME Test 1

Solution: If we create a table of values the quartic $f(x) = x^4 + 8x^3 + 15x^2 - 4x$ takes on for integers x, it appears that that it is symmetric about the line $x = -2$. Indeed, $f(-4-x) = (-4-x)^4 + 8(-4-x)^3 + 15(-4-x)^2 - 4(-4-x) = x^4 + 16x^3 + 96x^2 + 256x + 256 - 8x^3 - 96x^2 - 384x - 512 + 15x^2 + 120x + 240 + 4x + 16 = x^4 + 8x^3 + 15x^2 - 4x = f(x)$. Thus, the local maximum of $f(x)$ is at $x = -2$, so in order to have 4 distinct real roots, we must have that $f(-2) - n > 0$. $f(-2) = 20$, so the maximum value n can be is 19.

9. Let x, y, and z be positive real numbers such that

$$x + y + z = 21$$
$$\sqrt{x^2 + 16} + \sqrt{y^2 + 49} + \sqrt{z^2 + 81} = 29$$

If x can be expressed as $\dfrac{m}{n}$, where m and n are relatively prime positive integers, find $m + n$.

Solution: We can interpret this problem geometrically: Let $A = (0,0)$, $B = (x, 4)$, $C = (x+y, 11)$, and $D = (x+y+z, 20)$. The 1st equation implies that $AD = 29$, and the 2nd equation implies that $AB + BC + CD = 29 = AD$. The only way this happens is if the 4 points are collinear, meaning $20x = 21 * 4$ so $x = \frac{21}{5}$ giving an answer of 26.

10. Let $ABCD$ be a square and P be the point inside the square such that $PB = 10$, $PC = 8$, and $PD = 8$. The area of $ABCD$ can be written as $m + n\sqrt{p}$, where $m, n,$ and p are positive integers, and p is not divisible by the square of any prime. Find $m + n + p$.

Solution: Using the British flag theorem we see $PA = 10$. Let the side length of the square be $2k$, and drop altitude PH onto BC. We have that $PH = k$, $CH = a$, and $BH = 2k - a$ for some a. Then, note that $10^2 - (2k-a)^2 = 8^2 - a^2$ which simplifies to $4k^2 - 4ka = 36$, so $a = \frac{k^2 - 9}{k}$. Then, we plug this into $a^2 + k^2 = 64$ and multiplying both sides by k^2 we get the quartic $2k^4 - 82k^2 + 81 = 0$. Using the quadratic formula we get that $4k^2 = 82 \pm 14\sqrt{31}$. There are actually two different sizes that square $ABCD$ can be, but in the smaller size, P is outside the square, thus we have that the area of the square is $82 + 14\sqrt{31}$ giving a final answer of 127.

11. Let $\triangle ABC$ have side lengths $AB = 20$, $BC = 9$, and $CA = 15$ and P be an arbitrary point on side $\overline{BC}$. Let ω_1 and ω_2 be the inscribed circles of $\triangle ABP$ and $\triangle ACP$, respectively. The common exterior tangent of ω_1 and ω_2 (other than BC) intersects with AP at Q. Find AQ.

Solution: Let ω_1 and ω_2 have centers O_1 and O_2 respectively, let ω_1 and ω_2 be tangent to BC at points D and E respectively, let the common exterior tangent of ω_1 and ω_2 other than BC be tangent to ω_1 and ω_2 at points F and G respectively, and let ω_1 and ω_2 be tangent to AP at points H and I respectively. Observe the following:

- $DE = \sqrt{O_1 O_2^2 - (O_1 D - O_2 E)^2} = \sqrt{O_1 O_2^2 - (O_1 F - O_2 G)^2} = FG$
- $DE + FG = PD + PE + QF + QG = PH + HQ + PI + IQ = 2PQ$, thus $PQ = DE = FG$.
- Therefore $AQ = AP - PQ = AP - DE = AP - PD - PE$.

But note that $PD = \frac{AP+PB-AB}{2}$ and $PE = \frac{AP+PC-AC}{2}$ so

$$AP - PD - PE = \frac{AB + AC - BC}{2} = 13$$

12. Hamilton is standing on vertex A of regular hexagon $ABCDEF$. Every minute, he will walk to one of the two vertices adjacent to the one he is standing on. When he reaches point D, he will stop. In how many different ways can he reach point D after exactly 13 steps?

Solution: Let a_n be the number of ways to reach point A in n steps, let b_n be the number

Solutions of AIME Test 1

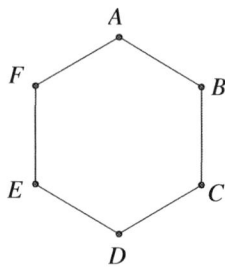

of ways to reach point B in n steps, and so on. So we are looking for d_{13}. We now compute a recursive formula for d_n:

$$d_n = e_{n-1} + c_{n-1} = f_{n-2} + b_{n-2} = e_{n-3} + 2a_{n-3} + c_{n-3} = 3f_{n-4} + 3b_{n-4} = 3d_{n-2}$$

And since $d_3 = 2$, we see that $d_{13} = 2 \cdot 3^5 = 486$.

13. A positive integer greater than 1 is called a *good number* if it can be divided by any positive integer less than its square root. How many good numbers are there?

Solution: We use bounding. Say that $k^2 < n \leq (k+1)^2$. Then clearly $k|n$, $(k-1)|n$, and $(k-2)|n$, so $\text{lcm}(k, k-1, k-2)|n$. So

$$\frac{(k)(k-1)(k-2)}{2} \leq \text{lcm}(k, k-1, k-2) \leq n \leq (k+1)^2$$

$$k(k-1)(k-2) \leq 2(k+1)^2$$

$$k^3 - 5k^2 - 2k - 2 \leq 0$$

It is easy to check that this condition fails for all $k \geq 6$, implying all good numbers do not exceed 36. From here, manual testing finds that the only good numbers are 2, 3, 4, 6, 8, 12, and 24, for a total of 7 good numbers.

14. Find the number of integers n such that $0 \leq n < 2021$ and 2021 divides $n^6 - 1$.

Solution: Notice that by the Chinese remainder theorem, we are just looking for the number of solutions $0 \leq n < 43$ to $n^6 \equiv 1 \pmod{43}$ multiplied by the number of solutions $0 \leq n < 47$ to $n^6 \equiv 1 \pmod{47}$. Now we use primitive roots. Let g be a primitive root of 43, then the solutions to $n^6 \equiv 1 \pmod{43}$ are in the form g^a for $0 < a \leq 42$ where $6a$ is a multiple of 42, meaning a is a multiple of 7, resulting in 6 solutions, $a = 7, 14, 21, 28, 35, 42$. Similarly, if h is a primitive root of 47, the solutions to $n^6 \equiv 1 \pmod{47}$ are in the form g^b for $0 < b \leq 46$ such that $6b$ is a multiple of 46, meaning b is a multiple of 23, resulting in 2 solutions, $b = 23, 46$. Thus, the total number of solutions is just $6 \cdot 2 = 12$.

15. In $\triangle ABC$, $\angle BAC = 120°$ and $AB : AC = 7 : 8$. Let P be an arbitrary point in $\triangle ABC$ on the angle bisector of $\angle BAC$. The minimum possible value of the ratio $\frac{BP}{CP}$ can be written as $\frac{\sqrt{m}}{n}$, where m and n are relatively prime positive integers. Find $m + n$.

Solution: Let $AP = x$, then by the law of cosines, we are trying to minimize
$$\sqrt{\frac{x^2 - 7x + 49}{x^2 - 8x + 64}}$$
which is the same as minimizing
$$\frac{x^2 - 7x + 49}{x^2 - 8x + 64} - 1 = \frac{x - 15}{x^2 - 8x + 64}$$
Now, substitute in $x = 15 - y$ (clearly $x < 15$), so we are trying to minimize
$$\frac{-y}{y^2 - 22y + 169}$$
Which is the same as maximizing
$$\frac{y^2 - 22y + 169}{-y} = 22 - (y + \frac{169}{y})$$
And by AM-GM, $y + \frac{169}{y}$ is minimized when $y = 13$, so $x = 2$, and thus the minimum value of $\frac{BP}{CP}$ is
$$\sqrt{\frac{4 - 14 + 49}{4 - 16 + 64}} = \sqrt{\frac{39}{52}} = \frac{\sqrt{3}}{2}$$
So our answer is 5.

Solutions of AIME Test 2

1. Real numbers x, y, and z satisfy the following inequality system:

$$1 \le 3x + y - z \le 11$$
$$-3 \le x - 3y + 5z \le 7$$
$$-4 \le x - 2y + z \le 6$$

If $x + y + z$ is an integer, find the sum of all possible values of $x + y + z$.

Solution: In solving this problem, we can use our knowledge of system of equations and basic inequality properties together. Let a, b, c be real numbers that

$$a(3x + y - z) + b(x - 3y + 5z) + c(x - 2y + z) = x + y + z.$$

Then,

$$3a + b + c = 1$$
$$a - 3b - 2c = 1$$
$$-a + 5b + c = 1$$

and we get $(a, b, c) = (1/2, 1/2, -1)$. If we multiply the given inequalities by 1, 1, -2 respectively, we will get $2x + 2y + 2z$. So if

$$1 \le 3x + y - z \le 11$$
$$-3 \le x - 3y + 5z \le 7$$
$$-12 \le -2x + 4y - 2z \le 8$$

and add up these inequalities, we get $-14 \le 2x + 2y + 2z \le 26$. Therefore,

$$-7 \le x + y + z \le 13.$$

The sum is $(-7) + (-6) + \cdots + 12 + 13 = \dfrac{6 \cdot 21}{2} = 63$.

Note: It is also possible to find triples (x, y, z) where $x + y + z$ equals -7 or 13. It will be helpful for the reader to review this part.

Solutions of AIME Test 2

2. In $\triangle ABC$, $AB = 187$ and $BC - CA = 119$. The length of the altitude from C to AB is 204, and $\cot \angle C = \dfrac{m}{n}$, where m and n are relatively prime positive integers. Find $m + n$.

   Solution: Let the foot of the altitude from C to AB be H. All of the numbers in the problem are multiples of 17, so we scale the problem down by a factor of 17 in order to make the numbers more manageable. Note that this doesn't change the value of $\cot \angle C$. $AB = 11$, $BC - CA = 7$, and $CH = 12$. We also need to make sure whether $\angle BAC$ is acute or obtuse. If it is an acute angle, note that $CA > 12$ and $BC < \sqrt{11^2 + 12^2}$, so $BC - CA < \sqrt{265} - 12 < 5$, contradiction, so $\angle BAC$ is obtuse. Now let $HA = x$. We have that

   $$\sqrt{(x+11)^2 + 12^2} - \sqrt{x^2 + 12^2} = 7$$

   $$(x+11)^2 + 12^2 = x^2 + 12^2 + 49 + 14\sqrt{x^2 + 12^2}$$

   $$11x + 36 = 7\sqrt{x^2 + 144}$$

   $$121x^2 + 792x + 1296 = 49x^2 + 7056$$

   $$72x^2 + 792x - 5760 = 72(x-5)(x+16) = 0$$

   Thus $x = 5$. Finally, note that $\tan(\angle C) = \tan(\angle BCH - \angle ACH) = \dfrac{\frac{11}{12}}{1 + \frac{80}{144}} = \dfrac{33}{56}$, so $\cot \angle C = \dfrac{56}{33}$ and the answer is 89.

3. When people buy tickets at a movie theater, their birthdays are recorded. A one-time free ticket is given to the first person in line who was born on the same day as some customer who bought a ticket before. Walking up to the ticket office, Andrew can enter the line wherever he wants. For example, he may choose to be the 1st person in line or the 1000th, or somewhere in between. Andrew does not know the birthday of anyone else in the line. In order to maximize his chances of getting the free ticket, Andrew queues as the nth person in the line. Find n.
   (Assume the odds of a person having a birthday on any of the 365 days in a year to be equal, and that no one has a birthday on leap day.)

Solutions of AIME Test 2

Solution: Let a_n be the probability that, if Andrew queues in the nth spot in line, he will get the free ticket. For example, $a_1 = 0$ as clearly the first person in line cannot get the free ticket. $a_2 = \frac{1}{365}$ as the 2nd person in line will get the ticket if they have the same birthday as the first person in line. $a_3 = \frac{364}{365} \cdot \frac{2}{365}$ as the 3rd person will get the ticket if the 2nd person doesn't have the same birthday as the first, and the 3rd person does have the same birthday as one of the first two people. We can recursively define this sequence: for all $k \geq 2$, we have that

$$a_{k+1} = a_k \cdot \frac{366-k}{365} \cdot \frac{k}{k-1}$$

and now notice that

$$a_{k+1} > a_k \iff k(366-k) > 365(k-1) \iff 365 > k^2 - k \iff k \leq 19$$

Thus, the sequence is strictly increasing until a_{20} and from there is strictly decreasing, so a_{20} is the highest value in the sequence, so for the highest probability Andrew would want to wait in the 20th spot.

4. Let $ABCD$ be a tetrahedron with $AB = AC = 3\sqrt{13}$, $BC = 18$, $AD = 10$, and $BD = CD = 11$. Find the volume of the tetrahedron.

Solution: Let F be midpoint of BC. Thus $BF = CF = 9$ and $DF^2 = 11^2 - 9^2 = 40$. Let's draw the perpendicular line from D to the plane of $\triangle ABC$. Let E be foot of the perpendicular line. Since, $AB = AC$ and $DB = DC$, point E lie on the line AF. $AF^2 = (3\sqrt{13})^2 - 9^2 = 36$ and $AF = 6$. Also, $AD^2 > AF^2 + DF^2$ and $\angle AFD$ is obtuse. Let $DE = h$ and $EF = x$. By Pythagoras' theorem

$$10^2 - (6+x)^2 = h^2 = 40 - x^2.$$

In these equations, we can find that $x = 2$, $h = 6$. $Area(ABC) = \frac{18 \cdot 6}{2} = 54$ and volume of the pyramid $V = \frac{1}{3} Area(ABC) \cdot h = 108$.

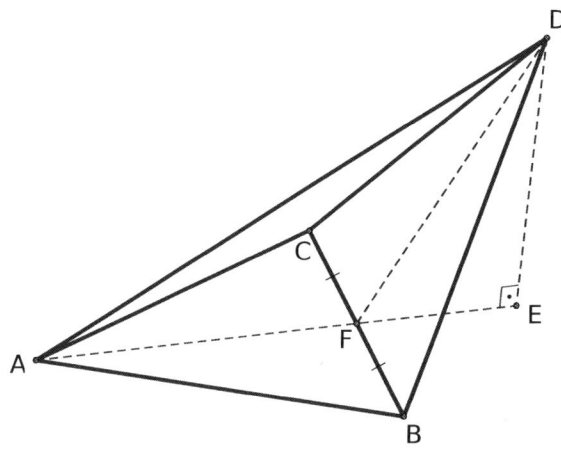

5. If $m = \sum_{n=1}^{300} \left\lfloor \dfrac{2^n}{5} \right\rfloor$, what is $2^{301} - 5m$?

($\lfloor x \rfloor$ equals the largest integer not greater than x.)

Solution: Notice that
$$2^n - 5 \left\lfloor \dfrac{2^n}{5} \right\rfloor = \begin{cases} 2 & n \equiv 1 \pmod{4} \\ 4 & n \equiv 2 \pmod{4} \\ 3 & n \equiv 3 \pmod{4} \\ 1 & n \equiv 0 \pmod{4} \end{cases}$$

Let $f(n) = 2^n - 5 \left\lfloor \dfrac{2^n}{5} \right\rfloor$. Then, it is clear from the above definition that $f(k) + f(k+1) + f(k+2) + f(k+3) = 10$, so
$$2^{301} - 5m = 2 + \sum_{n=1}^{300} f(n) = 2 + 10 \cdot 75 = 752$$

6. The solutions to the equation $z^{24} + 64z^{12} + 4096 = 0$ are the vertices of a convex polygon in

Solutions of AIME Test 2

the complex plane. The area of the polygon is $(\sqrt{a} - \sqrt{b}) \cdot \cos(5°)$, where a and b are positive integers. Find $a + b$.

Solution: Note this solution uses the *cis* function, which is defined as $\text{cis}(\theta°) = \cos(\theta°) + i\sin(\theta°)$. The equation is equivalent to

$$(z^{12} + 32)^2 = -3072$$

$$z^{12} = \pm 32i\sqrt{3} - 32$$

which splits up into the two equations $z^{12} = 64 \text{ cis } 120°$ and $z^{12} = 64 \text{ cis } 240°$. From here, we can compute all 24 solutions, which are $z = \sqrt{2} \text{ cis}(10° + 30°k)$ and $z = \sqrt{2} \text{ cis}(20° + 30°k)$, for k from 0 to 11. Let O be the origin of the complex plane. If we draw segment OP for each of the 24 points P, we see that the polygon is made up of 24 isosceles triangles with legs equal to $\sqrt{2}$ in length, 12 of the triangles having a vertex angle of 10 degrees and 12 having a vertex angle of 20 degrees. Thus, the total area of the polygon equals $12\sin(10°) + 12\sin(20°)$. Now we have to manipulate this value until we get the format the problem wants:

$$12\sin(10°) + 12\sin(20°) = 12 * 2\sin(5°)\cos(5°) + 12 * 2 * 2\sin(5°)\cos(5°) * \cos(10°)$$
$$= 24\cos(5°)[\sin(5°) + \sin(5°)\cos(10°) + \sin(5°)\cos(10°)]$$
$$= 24\cos(5°)[\sin(5°) + \sin(5°)(2\cos^2(5°) - 1) + \sin(5°)\cos(10°)]$$
$$= 24\cos(5°)[2\sin(5°)\cos^2(5°) + \sin(5°)\cos(10°)]$$
$$= 24\cos(5°)[\sin(10°)\cos(5°) + \sin(5°)\cos(10°)]$$
$$= 24\cos(5°)\sin(15°)$$
$$= \cos(5°)(6\sqrt{6} - 6\sqrt{2})$$
$$= \cos(5°)(\sqrt{216} - \sqrt{72})$$

Thus our answer is $216 + 72 = 288$.

Note: There is an easier way to do this manipulation if you know the following identity:
$$\sin(3x) = 3\sin(x) - 4\sin^3(x)$$

Solutions of AIME Test 2

7. Positive integers $a, b,$ and c satisfy the equation

$$ab(b^3 - a^3) + bc(c^3 - b^3) + ca(a^3 - c^3) + a^2b^2(b-a) + b^2c^2(c-b) + c^2a^2(a-c) = 450.$$

Find abc.

Solution: Let $F(a,b,c) = ab(b^3 - a^3) + bc(c^3 - b^3) + ca(a^3 - c^3) + a^2b^2(b-a) + b^2c^2(c-b) + c^2a^2(a-c)$. We see that F is cyclic for a, b, c. For $b = c$ $F(a, b, b) = 0$ thus $(b - c) \mid F$. Similarly $(c - a) \mid F$ and $(a - b) \mid F$. Since F fifth degree homogeneous,

$$F(a, b, c) = (a-b)(b-c)(c-a) \left[A(a^2 + b^2 + c^2) + B(ab + bc + ca) \right],$$

where A, B are real numbers. We can find that $A = 1, B = 2$ with using some special values of a, b, c. Then $F(a, b, c) = (a-b)(b-c)(c-a)(a+b+c)^2$ and we have to solve the equation

$$(a-b)(b-c)(c-a)(a+b+c)^2 = 450$$

over positive integers. We see that a, b, c are different positive integers. Hence $a+b+c \geq 6$. Since $450 = 2 \cdot 3^2 \cdot 5^2$, $a+b+c$ only can be 15. Then $(a-b)(b-c)(c-a) = 2$ and all solutions in the forms $(a, b, c) = (n, n+1, n+2), (n+1, n+2, n), (n+2, n, n+1)$. On the other hand, $a+b+c = 3n+3 = 15$ and $n = 4$. Thus, $abc = 4 \cdot 5 \cdot 6 = 120$.

8. Let $S = \binom{40}{0} + \binom{40}{4} + \binom{40}{8} + \cdots + \binom{40}{40}$. Find the remainder when S is divided by 1000.

Solution: Let $f(x) = (x+1)^{40}$. Now, we note that:

$$f(1) = \binom{40}{0} + \binom{40}{1} + \cdots + \binom{40}{40}$$

$$f(-1) = \binom{40}{0} - \binom{40}{1} + \binom{40}{2} - \cdots + \binom{40}{40}$$

$$f(i) = \binom{40}{0} + i\binom{40}{1} - \binom{40}{2} - i\binom{40}{3} + \cdots - i\binom{40}{39} + \binom{40}{40}$$

Solutions of AIME Test 2

$$f(-i) = \binom{40}{0} - i\binom{40}{1} - \binom{40}{2} + i\binom{40}{3} + \cdots + i\binom{40}{39} + \binom{40}{40}$$

$$f(1) + f(-1) + f(i) + f(-i) = 4S$$

$$S = \frac{2^{40} + (1+i)^{40} + (1-i)^{40}}{4} = \frac{2^{40} + 2^{20} + 2^{20}}{4} = 2^{38} + 2^{19} = 2^{19}(2^{19} + 1)$$

and now we compute by hand that $2^{19} \equiv 288 \pmod{1000}$ so $S \equiv 288*289 \equiv 232 \pmod{1000}$.

9. a, b, c positive numbers and $a + b + c = 6$. Minimum value of

$$\frac{\log_a b}{2a - 3b + 20} + \frac{\log_b c}{b - c + 18} + \frac{\log_c a}{28 - c - 4a}$$

is $\frac{m}{n}$, where m and n are relatively coprime positive integers. Find $m + n$.

Solution: Let's apply arithmetic mean-geometric mean inequality twice for

$$S = \frac{\log_a b}{2a - 3b + 20} + \frac{\log_b c}{b - c + 18} + \frac{\log_c a}{28 - c - 4a}.$$

$$S \geq 3\sqrt[3]{\frac{\log_a b}{2a - 3b + 20} \cdot \frac{\log_b c}{b - c + 18} \cdot \frac{\log_c a}{28 - c - 4a}}$$

$$= \frac{3}{\sqrt[3]{(2a - 3b + 20) \cdot (b - c + 18) \cdot (28 - c - 4a)}}$$

$$\geq \frac{3 \cdot 3}{(2a - 3b + 20) + (b - c + 18) + (28 - c - 4a)}.$$

Thus we find, $S \geq \frac{9}{66 - 2(a + b + c)}$. Since $a + b + c = 6$, we yields $S \geq \frac{1}{6}$. The equality condition is satisfy when $a = b = c = 2$. Therefore, minimum value of S is $\frac{1}{6}$ and we get $m + n = 1 + 6 = 7$.

10. What is the remainder of the product of n integers satisfying the conditions $1 \leq n \leq 1000$ and $\gcd(n, 1000) = 1$, divided by 1000?

Solution: Pair every integer a with its corresponding multiplicative inverse b (the unique

residue such that $ab \equiv 1 \pmod{1000}$), if it is distinct from a. This eliminates every residue except the residues x for which $x^2 \equiv 1 \pmod{1000}$. This happens when $(x-1)(x+1)$ is a multiple of 1000. It is a multiple of 8 whenever x is odd, and it is a multiple of 125 whenever x is 1 more or less than a multiple of 125. Checking, the values that correspond to these conditions are $x = 1, 249, 251, 499, 501, 749, 751, 999$. Our answer is just the product of these values mod 1000, which is just

$$1 \cdot 249 \cdot 251 \cdot 499 \cdot 501 \cdot 749 \cdot 751 \cdot 999 \equiv 1 \cdot 249 \cdot 251 \cdot 499 \cdot 499 \cdot 251 \cdot 249 \cdot 1 \equiv 1 \pmod{1000}$$

11. A, B, C, D, and E lie on a circle in that order. $\angle ABE = \angle DBC$ and $\angle BAC = \angle DAE$. BC and AE intersect at F. $AB = 6$, $CE = 5$, and the circumradius of triangle ABF is $\frac{m}{n}$, where m and n are relatively prime positive integers. Find $m + n$.

Solution: Let $\angle BAC = \angle DAE = \alpha$, $\angle ABE = \angle DBC = \beta$. By inscribed angles, $\angle EAC = \angle ECD = \alpha$ and $\angle ACE = \angle DEC = \beta$. Thus, we conclude that $EB \parallel CD$, $AC \parallel ED$, $BCDE$ and $ACDE$ are isosceles quadrilaterals. Then $BD = CE = AD = 5$, $\angle EAB = 2\alpha + \beta$, $\angle CBA = 2\beta + \alpha$. Let H be the foot of the perpendicular drawn from point D to AB. So $AH = HB = 3$ and $DH = 4$. Let $\theta = \alpha + \beta$. Since $\angle FAB + \angle FBA = (2\alpha + \beta) + (2\beta + \alpha) = 3\theta$, we get $\angle AFB = 180° - 3\theta$. We need to calculate $\sin 3\theta$. In the right triangle ADH, $\sin \theta = \frac{4}{5}$, $\cos \theta = \frac{3}{5}$. Then, $\sin 2\theta = 2\sin\theta\cos\theta = \frac{24}{25}$, $\cos 2\theta = \cos^2\theta - \sin^2\theta = -\frac{7}{25}$. $\sin 3\theta = \sin 2\theta \cos \theta + \cos 2\theta \sin \theta = \frac{44}{125}$. That is, $\sin \angle AFB = \sin 3\theta = \frac{44}{125}$. By applying the sine theorem in triangle AFB, $R = \frac{AB}{2\sin 3\theta}$ where R is circumradius of the triangle. Then, $R = \frac{6}{88/125} = \frac{375}{44}$ and $m + n = 375 + 44 = 419$.

Solutions of AIME Test 2

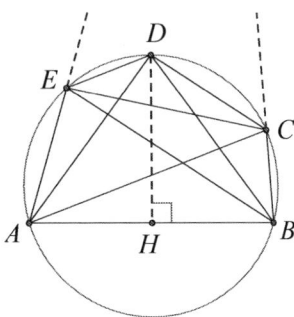

12. The polynomial $P(x) = x^3 + (2n-5)x^2 + (n^2+4n+3)x + n+3$ has three positive real roots given as x_1, x_2, x_3. If $\arctan(x_1) + \arctan(x_2) + \arctan(x_3) = \dfrac{3\pi}{4}$, what is the absolute value of the product of the coefficients of $P(x)$?

Solution: Let $\arctan(x_k) = \theta_k$, where $k = 1, 2, 3$. Given that $\theta_1 + \theta_2 + \theta_3 = \dfrac{3\pi}{4}$. We know that the identity

$$\tan(\theta_1 + \theta_2) = \frac{\tan(\theta_1) + \tan(\theta_2)}{1 - \tan(\theta_1)\tan(\theta_2)}.$$

Hence we get the identity,

$$\tan(\theta_1 + \theta_2 + \theta_3) = \frac{\tan(\theta_1) + \tan(\theta_2) + \tan(\theta_3) - \tan(\theta_1)\tan(\theta_2)\tan(\theta_3)}{1 - \tan(\theta_1)\tan(\theta_2) - \tan(\theta_1)\tan(\theta_3) - \tan(\theta_2)\tan(\theta_3)}.$$

Let $P(x) = x^3 + Ax^2 + Bx + C$, where $A = 2n-5$, $B = n^2+4n+3$, $C = n+3$. By Vieta's formula

$$\tan(\theta_1 + \theta_2 + \theta_3) = \frac{-A + C}{1 - B}.$$

Since $\tan(\theta_1+\theta_2+\theta_3) = \tan\dfrac{3\pi}{4} = -1$, we find that $C = A+B-1$. This gives $n+3 = n^2+6n-3$ and therefore, $n = 1$ or $n = -6$. Given that $x_k > 0$; thus $A < 0$, $B > 0$, $C < 0$. For $n = 1$, $C = 4$ is not a negative number. Hence $n = -6$ and $|A \cdot B \cdot C| = 17 \cdot 15 \cdot 3 = 765$.

13. If the n-th root of a positive integer with n digits in the decimal number system is n less than the sum of the digits of this number, the number is called a *wonderful number* where $n \geq 2$. For example, $\sqrt[3]{216} = 2 + 1 + 6 - 3$ and 216 is a wonderful number. What is the remainder of the sum of the wonderful numbers less than 10000 divided by 1000?

Solution: Let $abcd$ be a 4 digit number. Then $\sqrt[4]{abcd} = a + b + c + d - 4$ and

$$abcd = (a + b + c + d - 4)^4 \qquad (1)$$

Thus $abcd$ may be $6^4, 7^4 = 2401, 8^4 = 4096, 9^4$. If $abcd = 6^4$ or 9^4 then, $9 \mid abcd$ but $9 \nmid a + b + c + d - 4$. In addition, if the values of $2401, 4096$ for $abcd$ are examined, it is seen that equation (1) is not satisfied.

Let abc be a 3 digit number. Then

$$abc = (a + b + c - 3)^3 \qquad (2)$$

Thus abc may be $5^3 = 125, 6^3 = 216, 7^3 = 343, 8^3 = 512, 9^3 = 729$. If we check these values, we see that the numbers $125, 216, 343$ satisfy equation (2).

Let ab be a 2 digit number. Then

$$ab = (a + b - 2)^2 \qquad (3)$$

Thus abc may be $16, 25, 36, 49, 64, 81$. If we check these values, we see that the numbers $25, 64$ satisfy equation (3).

So the sum of all wonderful numbers less than 10000 is $25 + 64 + 125 + 216 + 343 = 773$. The remainder of this number divided by 1000 is itself.

14. A 16-person chess tournament will be held, with two students from 8 schools each participating. 8 simultaneous chess matches can be arranged in N different ways so that students

Solutions of AIME Test 2

from the same school do not match with each other. What is the remainder after dividing N by 1000?

Solution: Let the number of schools be n and the number of students be $2n$. Let $a(n)$ be the number of matches with the desired feature. Easily, we can find that $a(1) = 0$ and $a(2) = 2$. Let's denote students from the same school $\{A_i, B_i\}$ where $i = 1, 2, \ldots, n$. For $n > 2$, there are $2n - 2$ students that the student A_1 can match. Let's assume that A_1 and A_2 match. If B_1 and B_2 match, the remaining students $a(n-2)$ match in a different way. If B_1 and B_2 don't match, there are $2n - 4$ students that the student B_1 can match. In this case, we continue by examining the matches that B_2 can make, and we find that

$$a(n) = (2n-2)\left[a(n-2) + (2n-4)\left(a(n-3) + (2n-6)(a(n-4) + \cdots)\right)\right].$$

On the other hand we can write that

$$a(n-1) = (2n-4)\left(a(n-3) + (2n-6)(a(n-4) + \cdots)\right).$$

Hence, we yields that

$$a(n) = (2n-2)\left(a(n-1) + a(n-2)\right)$$

for $n > 2$. According to this recurrence relation, we get $a(3) = 8, a(4) = 60, a(5) = 544, a(6) = 6040$. Also $a(6) \equiv 40 \pmod{1000}$, $a(7) \equiv 12 \cdot (40 + 544) \equiv 8 \pmod{1000}$, $a(8) \equiv 14 \cdot (8 + 40) \equiv 672 \pmod{1000}$. So, the remainder of $a(8)$ divided by 1000 is 672.

15. $\triangle ABC$ has the length of sides $AB = 5$, $BC = 8$, $CA = 7$ and a point P is taken on the plane of $\triangle ABC$. Let $PA = x$, $PB = y$, $PC = z$. What is the minimum integer value of $5x + 7y + 8z$?

Solution: By cosine theorem, $\cos \angle ABC = \dfrac{8^2 + 5^2 - 7^2}{2 \cdot 8 \cdot 5} = \dfrac{1}{2}$. Therefore $\angle ABC = 60°$. Let's draw similar triangles $\triangle ABC \sim \triangle CBE$. Similarity ratio is $k = \dfrac{BC}{BA} = \dfrac{8}{5}$. We conclude that $\triangle CBE$ is a dilation of $\triangle ABC$. Let point D be image of point P under the dilation. Thus $BE = \dfrac{64}{5}$, $BD = \dfrac{8y}{5}$, $CD = \dfrac{8x}{5}$, $ED = \dfrac{8z}{5}$ and $\angle CBE = 60°$.

Solutions of AIME Test 2

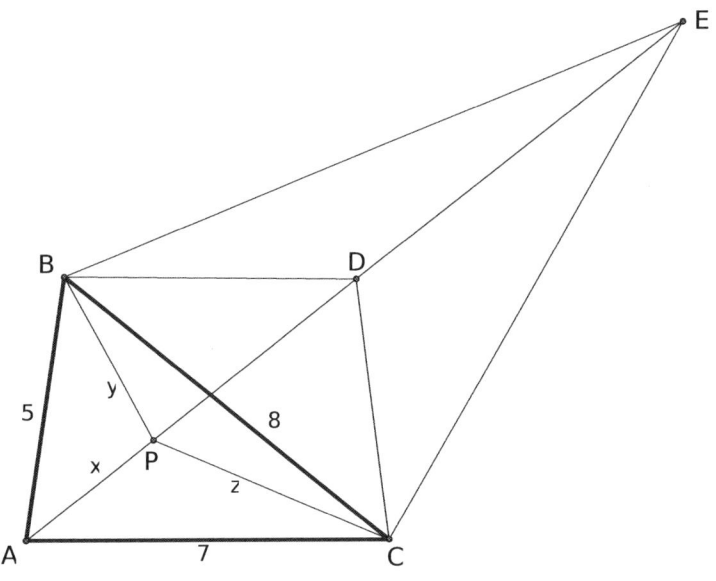

By triangle inequality, $AE \leq AP + PD + DE$ and $AE \leq x + \dfrac{7y}{5} + \dfrac{8z}{5}$. Then,

$$5x + 7y + 8z \geq 5 \cdot AE.$$

Let's apply cosine theorem in the $\triangle ABE$, $AE^2 = 5^2 + \left(\dfrac{64}{5}\right)^2 - 2 \cdot 5 \cdot \dfrac{64}{5} \cdot \cos 120°$. Hence, $AE = \dfrac{1}{5}\sqrt{6321}$ and

$$5x + 7y + 8z \geq \sqrt{6321}.$$

The equality condition is satisfy when the points A, P, D, E are collinear. Since $79^2 = 6241 < 6321 < 6400 = 80^2$, minimum integer value of $5x + 7y + 8z$ is 80.

Solutions of AIME Test 3

1. The table below shows how many hours it took Geralt of Rivia, Vesemir, Cirilla and Yennefer to complete a task as two people each.

	Cirilla	Yennefer
Geralt of Rivia	m/n	15
Vesemir	8	10

   Geralt of Rivia and Cirilla can complete the same task in m/n hours, whose m, n are coprime positive integers. What is $2m + n$?

   Solution: Let Geralt of Rivia, Vesemir, Cirilla, and Yennefer complete the task in a, b, c, d hours, respectively, when they are alone. We can write the equations
   $$\frac{1}{a} + \frac{1}{c} = \frac{n}{m}, \quad \frac{1}{b} + \frac{1}{d} = \frac{1}{10}.$$
   Similarly,
   $$\frac{1}{a} + \frac{1}{d} = \frac{1}{15}, \quad \frac{1}{b} + \frac{1}{c} = \frac{1}{8}.$$
   Thus,
   $$\frac{n}{m} + \frac{1}{10} = \frac{1}{15} + \frac{1}{8}$$
   and we find $\dfrac{n}{m} = \dfrac{11}{120}$. So, $2m + n = 251$.

2. Points $A(-2, 5)$, $B(2, 1)$, $C(8, 1)$ are given on the analytical plane. If $O(a, b)$, $H(c, d)$, $G(e, f)$ are circumcenter, orthocenter, centroid of the triangle ABC, respectively, what is $a^2 + b^2 + c \cdot d + e + f$?

   Solution: Slope of AB is $m_{AB} = \dfrac{5-1}{-2-2} = -1$. Then $\angle ABC = 135°$. Therefore, easily we can find $O(5, 8)$ and $H(-2, -9)$. Also, $e = \dfrac{2 + 8 + (-2)}{3} = \dfrac{8}{3}$ and $f = \dfrac{1 + 1 + 5}{3} = \dfrac{7}{3}$. Hence, $a^2 + b^2 + c \cdot d + e + f = 25 + 64 + 18 + 5 = 112$.

Solutions of AIME Test 3

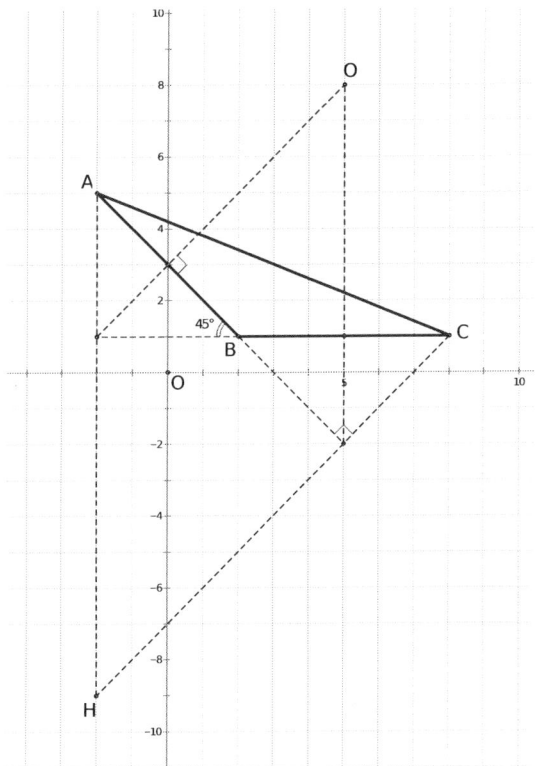

3. Let α, β, γ be roots of the equation $x^3 - 6x^2 - 331x - 8 = 0$. What is $\sqrt[3]{\alpha} + \sqrt[3]{\beta} + \sqrt[3]{\gamma}$?

Solution: Using the intermediate value theorem, we can see that α, β, γ are real numbers.
$(x-2)^3 = 343x$ and $x - 2 = 7\sqrt[3]{x}$. Thus, $\sqrt[3]{\alpha} + \sqrt[3]{\beta} + \sqrt[3]{\gamma} = \frac{1}{7}(\alpha - 2 + \beta - 2 + \gamma - 2)$. By Vieta's theorem, $\alpha + \beta + \gamma = 6$. So, we find $\sqrt[3]{\alpha} + \sqrt[3]{\beta} + \sqrt[3]{\gamma} = 0$.

4. a, b, c are positive integers satisfy the following equation

$$\frac{1}{a} + \frac{1}{b} + \frac{1}{c} = \frac{7}{24}.$$

What is the maximum value of $a + b + c$?

Solution: Let's assume that $c \leq b \leq a$. Since $\frac{1}{c} < \frac{7}{24}$, $c \geq 4$. For $c = 4$, we find that $\frac{1}{a} + \frac{1}{a} = \frac{1}{24}$. Thus $b \geq 25$. For $b = 25$, we find $a = 600$. Then, we get the maximum value $a + b + c = 600 + 25 + 4 = 629$.

Note that for $c = 4, b = 26$ we find $a = 312$. That is, increasing b by 1 makes the value of a much smaller. For this reason, we choose b and c as small as possible.

5. $ABCD$ is a square and it's divided into 9 equal squares. When going from corner A to corner C, only east, north or northeast directions can be used. At each point, the direction to go is chosen randomly with equal probability. For example, when at point B, the probability of going north is 1, since it is only possible to move in the north direction. The probability of going through point X on the way from A to C is m/n, whose m and n are co-prime positive integers. What is $m + n$?

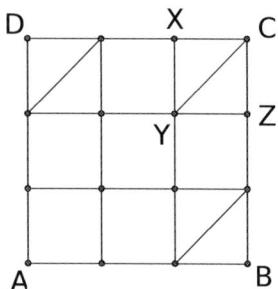

Solution: Let p_X be the probability of passing through point X on the way from A to C. By symmetry, $p_Z = p_X$. Also, $p_X + p_Y + p_Z = 1$. There are 6 different ways from point A to point Y. The probability of going from A to Y is

$$2\left(\frac{1}{2} \cdot 1 \cdot \frac{1}{3} \cdot \frac{1}{2} + \frac{1}{2} \cdot \frac{1}{2} \cdot \frac{1}{2} \cdot \frac{1}{2} + \frac{1}{2} \cdot \frac{1}{2} \cdot \frac{1}{2} \cdot \frac{1}{2}\right) = \frac{5}{12}.$$

Hence, $p_Y = \frac{1}{3} \cdot \frac{5}{12} = \frac{5}{36}$ and $p_X = p_Z = \frac{1 - 5/36}{2} = \frac{31}{72}$. Thus, $m + n = 31 + 72 = 103$.

6. $ABCDEF$ is a regular hexagon and P is a point inside of the hexagon. The areas of triangles APF, BPA, CPB are 27, 24, 15, respectively. What is the area of the hexagon?

Solution: Let M be intersection of AB and CD. BCM is an equilateral triangle.

$$Area(MCP) = Area(CDP) = 15, \quad Area(MBP) = Area(ABP) = 27.$$

Thus, $Area(MCPB) = 15 + 27 = 42$.

$$Area(MCB) = Area(MCPB) - Area(CPB) = 42 - 24 = 18.$$

We know $Area(ABCDEF) = 6 \cdot Area(MCB)$. So, $Area(ABCDEF) = 6 \cdot 18 = 108$.

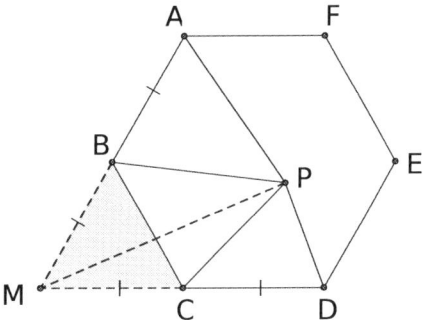

7. The Paradis Island scout unit consists of 1700 people. 91% of the scouts like Rogue Titan, 79% like Armored Titan, 77% like Colossal Titan, 86% like Female Titan. If the minimum and the maximum number of people in the scout unit that likes all four titans are a and b, respectively, then what is $b - a$?

Solution: Let the percentage of people who like X Titan be $P(X)$.
$100 \geq P(R \cup A) = P(R) + P(A) - P(R \cap A)$ and $P(R \cap A) \geq 70$. Similarly,
$100 \geq P(C \cup F) = P(C) + P(F) - P(C \cap F)$ and $P(C \cap F) \geq 63$. Also,

$$100 \geq P((R \cap A) \cup (C \cap F)) = P(R \cap A) + P(C \cap F) - P(R \cap A \cap C \cap F)$$

and we find that $P(R \cap A \cap C \cap F) \geq 70 + 63 - 100 = 33$. There are at least $a = 1700 \cdot \dfrac{33}{100}$ people. Also, $P(R \cap A \cap C \cap F) \leq P(C) = 77$. Equality is achieved when the set of those who like the Colossal Titan is a subset of each of those who like the other Titans. There are at most $b = 1700 \cdot \dfrac{77}{100}$ people. So, $b - a = 1700 \cdot \dfrac{44}{100} = 748$.

Solutions of AIME Test 3

8. The f_n Fibonacci sequence is defined as $f_1 = f_2 = 1$ and for $n \geq 1$, $f_{n+2} = f_{n+1} + f_n$.

$$\sum_{n=1}^{14} \left(\frac{2n}{f_n} - \frac{n \cdot f_{n+2}}{f_{n+1} \cdot f_n} - \frac{1}{f_{n+1}} \right) = \frac{m}{n},$$

where m and n are co-prime positive integers. What is $m + n$?

Solution: $\dfrac{2n}{f_n} - \dfrac{n \cdot f_{n+2}}{f_{n+1} \cdot f_n} - \dfrac{1}{f_{n+1}} = \dfrac{2n \cdot f_{n+1} - n(f_{n+1} + f_n) - f_n}{f_{n+1} \cdot f_n} = \dfrac{n \cdot f_{n+1} - (n+1) \cdot f_n}{f_{n+1} \cdot f_n}$

$= \dfrac{n}{f_n} - \dfrac{n+1}{f_{n+1}}$. It's a telescopic expression, therefore

$$\sum_{n=1}^{14} \left(\frac{n}{f_n} - \frac{n+1}{f_{n+1}} \right) = \frac{1}{f_1} - \frac{15}{f_{15}}.$$

Easily we can calculate that $f_{15} = 610$. Thus, $1 - \dfrac{15}{610} = \dfrac{119}{122}$ and $m + n = 119 + 122 = 241$.

9. In the decimal system, if the rightmost digit of a positive integer is replaced by the leftmost digit, the resulting number is equal to 3 times the first number. What is the remainder of the division by 1000 of the smallest number satisfying this condition?

Solution: Let $m = abxy \ldots z$ be our minimum number. Let $n + 1$ be number of digits m. Given condition is $bcxy \ldots za = 3 \cdot m$ and so, $10 \cdot bcxy \ldots z + a = 3 \cdot (a \cdot 10^n + bxy \ldots z)$. Then,

$$\frac{a}{7}(3 \cdot 10^n - 1) = bxy \ldots z < 10^n.$$

If $a \geq 3$, left-hand side greater than right-hand side. Thus $a \in \{1, 2\}$. Also, $3 \cdot 10^n - 1 \equiv 0 \pmod{7}$. We can find that $3^n \equiv 5 \pmod 7$. Minimum n is 5. Since $\text{ord}_7(3) = 6$, $n = 6k + 5$, (k is a non-negative integer). If we take $a = 1$, $bxy \ldots z = \dfrac{1}{7}(3 \cdot 10^5 - 1) = 42857$ and $m = 142857$. Therefore $m \equiv 857 \pmod{1000}$.

10. The side lengths of triangle ABC are integers and $\angle BAC = \alpha$, $\angle ABC = \beta$, $\angle BCA = \gamma$. If $\sin \beta = 4 \sin \alpha \cos \gamma$, what is the minimum value of the perimeter of triangle ABC?

Solutions of AIME Test 3

Solution: Let $BC = a$, $CA = b$, $AB = c$. By sine rule, $\dfrac{b}{a} = \dfrac{\sin\beta}{\sin\alpha} = 4\cos\gamma$. By cosine theorem, $\cos\gamma = \dfrac{a^2 + b^2 - c^2}{2ab}$. Hence, we get

$$b^2 = 2(c^2 - a^2).$$

Thus, b is an even integer and $a < c$. By triangle inequality $a > |c - b| \geq 1$, then $a \geq 2$. For the minimum value, we assume that a, b, c are co-prime integers. We see that a and c have same parity. If a and c are even then, a, b, c aren't co-prime. Therefore a and c are odd co-prime integers. $a \geq 3$ and $c \geq 5$. Let's make a table:

c	a	b
5	3	No
7	3	No
7	5	No
9	5	No
9	7	8
11	3	No

Then, minimum perimeter is $a + b + c = 7 + 8 + 9 = 24$.

Together with these; in the equation $b^2 = 2(c^2 - a^2)$, we get that $4 \mid b$. When $b = 2(2n+1)$, (n is an integer) we find $c^2 - a^2 = 2(2n+1)^2$. We know that this equation gives a contradiction in modulo 4. Therefore $4 \mid b$. For $b = 4$, $c^2 - a^2 = 8$ and $c = 3, a = 1$. But $a \geq 2$, it's impossible. Then $b \geq 8$.

11. A regular pentagonal pyramid is a solid with a regular pentagonal base and equilateral triangular lateral surfaces. The volume of a regular pentagonal pyramid with a side length of 6 is $m + n\sqrt{p}$. Here m, n, p are positive integers and p is a square-free number. What is $m+n+p$?

Solution: We know that $\sin 54° = \dfrac{\sqrt{5}+1}{4}$. Then, $\tan^2 54° = \dfrac{\sin^2 54°}{1 - \sin^2 54°} = \dfrac{5 + 2\sqrt{5}}{5}$.

67

Solutions of AIME Test 3

Let a, r, h be a side length, inradius of the regular pentagon, height of the regular pentagonal pyramid, respectively. Length of an altitude of equilateral triangle is $h' = \frac{a\sqrt{3}}{2}$. $r = \frac{a}{2}\tan 54°$. Since $(h')^2 = r^2 + h^2$, we find $h = \frac{a}{2}\sqrt{3 - \tan^2 54°} = \frac{a}{2}\sqrt{\frac{10 - 2\sqrt{5}}{5}}$. Therefore, volume of the regular pentagonal pyramid is

$$V = \frac{1}{3} \cdot \left(5 \cdot \frac{ar}{4}\right) \cdot h = \left(\frac{5 + \sqrt{5}}{24}\right) a^3.$$

For $a = 6$, we get $V = 45 + 9\sqrt{5}$. So, $m + n + p = 45 + 9 + 5 = 59$.

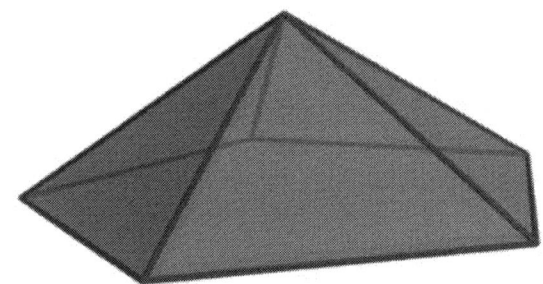

12. A function f defined on real numbers satisfy the equation

$$f(x + y) + f(x - y) = 2f(x) + 2f(y) - 6y$$

for every real number x, y. If $f(1) = 4$, what is $f(-24)$?

Solution: For $x = y = 0$, we get $f(0) = 0$. For $y = x$, we get $f(2x) = 4f(x) - 6x$ for all x real numbers. For $x = 0$, we get $f(y) = f(-y) + 6y$ and so $f(x) = f(-x) + 6x$ for all x real numbers. For $y = 2x$, we get $f(3x) + f(-x) = 2f(x) + 2f(2x) - 12x$ and thus $f(3x) = 9f(x) - 18x$ for all x real numbers. Therefore,

$f(2) = 4f(1) - 6 \cdot 1$ and $f(2) = 10$.
$f(4) = 4f(2) - 6 \cdot 2$ and $f(4) = 28$.
$f(8) = 4f(4) - 6 \cdot 4$ and $f(8) = 88$.
$f(24) = 9f(8) - 18 \cdot 8$ and $f(24) = 648$.
$f(24) = f(-24) + 6 \cdot 24$ and $f(-24) = 504$.

Solutions of AIME Test 3

We can also search for a quadratic function satisfying all the given conditions. There is at least one such function, $f(x) = x^2 + 3x$.

13. There are 1000 cities in a country. There are 250000 roads in total between cities. If there are direct roads from one city to all other cities, we call that city *metropolis*. What is the maximum number of metropolises?

    Solution: If there are n metropolises, it becomes $1000+999+998+\cdots+(1001-n) \leq 250000$ which $n \leq 1000$. Also, $n \geq 250$. Let $f(n) = 1000 + 999 + 998 + \cdots + (1001 - n)$. That is, $f(n) = (2001 - n)n/2$. We have to find largest n integer that $f(n) \leq 250000$. We can see that f is monotonic increasing on the interval $250 \leq n \leq 1000$. Let $g(n) = 250000 - f(n)$.

    For $n = 250$, $g(250) = 31125$. Let's increase n by 31.

    For $n = 281$, $g(281) = 8340$. Let's increase n by 8.

    For $n = 289$, $g(289) = 2616$. Let's increase n by 2.

    For $n = 291$, $g(291) = 1195$. Let's increase n by 1.

    For $n = 292$, $g(292) = 486$. Let's increase n by 1.

    For $n = 293$, $g(293) = -222 < 0$. Therefore $n < 293$ and we conclude that maximum value of n is 292.

14. P is an arbitrary point on the circumcircle of the equilateral triangle ABC. If radius of the circumcircle is $\sqrt{6}$, what is $PA^4 + PB^4 + PC^4$?

    Solution: Let r be radius of the circumcircle and $BC = a, PB = x, PA = y, PC = z$. We will prove that $PA^4 + PB^4 + PC^4 = 2a^4$.

    Without loss of generality, let's assume that P is on minor arc $\overparen{BC}$. By Ptolemy theorem,

Solutions of AIME Test 3

$y = x + z$. From cosine theorem;

$$a^2 = x^2 + z^2 + xz = x^2 + y^2 - xy = z^2 + y^2 - yz.$$

If we use the perfect square identity $y^2 = (x+z)^2 = x^2 + z^2 + 2xz$ in $a^2 = x^2 + z^2 + xz$ then, $y^2 = a^2 + xz$. By summing the equalities $a^2 = x^2 + y^2 - xy$ and $a^2 = z^2 + y^2 - yz$, we find $2a^2 = x^2 + 2y^2 + z^2 - y(x+z)$. Since $y = x + z$, we get $x^2 + y^2 + z^2 = 2a^2$. Now we will show $x^2y^2 + y^2z^2 + x^2z^2 = a^4$.

$x^2y^2 + y^2z^2 + x^2z^2 = (x^2 + z^2)y^2 + x^2z^2 = (a^2 - xz)y^2 + x^2z^2 = a^2y^2 - xzy^2 + x^2z^2 = a^2y^2 - xz(y^2 - xz) = a^2y^2 - xza^2 = a^2(y^2 - xz) = a^2 \cdot a^2 = a^4$. By the perfect square identity $(x^2 + y^2 + z^2)^2 = x^4 + y^4 + z^4 + 2(x^2y^2 + y^2z^2 + x^2z^2)$, we find that

$$x^4 + y^4 + z^4 = 2a^4.$$

Especially; for $r = \sqrt{6}$, $a = 3\sqrt{2}$. We get $PA^4 + PB^4 + PC^4 = 2 \cdot (3\sqrt{2})^4 = 648$.

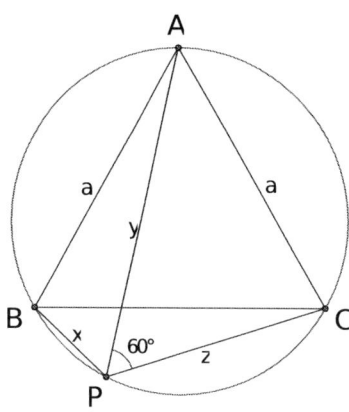

15. p and $p^4 - 35p^3 + 365p^2 - 1225p + 1259$ are prime numbers. What is sum of all possible values of such p primes?

Solution: Let $f(p) = p^4 - 35p^3 + 365p^2 - 1225p + 1259$. For $p = 5$, $f(5) = 509$ is a prime number. Let $p \neq 5$. Then, by Fermat's theorem $p^4 \equiv 1 \pmod{5}$. Therefore,

$$f(p) = p^4 - 35p^3 + 365p^2 - 1225p + 1259 \equiv 1 + 4 \equiv 0 \pmod{5}$$

and $5 \mid f(p)$. Aslo, $f(p)$ is a prime number. Hence, $f(p)$ have to equal 5. $p^4 - 35p^3 + 365p^2 - 1225p + 1259 = 5$ and we find the equation

$$(p-2)(p-3)(p-11)(p-19) = 0.$$

Thus, we find four more solutions for p: $2, 3, 11, 19$. So, $p \in \{2, 3, 5, 11, 19\}$ and sum of these is 40.

Solutions of AIME Test 4

1. Let $s(n)$ denote the sum of the digits of n, when written in base 10. For example, $s(123) = 6$ and $s(2024) = 8$. Let X be the number of positive integers n less than 10^5 that have the property that $s(n) > 10$. Find the first 3 digits of X.

Solution: We use complementary counting, subtracting the number of positive integers satisfying $s(n) \leq 10$ from 99999. Using stars and bars (10 stars, 5 bars) we get $\binom{15}{5}$ possiblities. This overcounts by 6, though, as the numbers 00000, (10)0000, 0(10)000, 00(10)00, 000(10)0, and 0000(10) are counted. So in total our answer is $99999 - \binom{15}{5} + 6 = 97002$ making our final answer 970.

2. From point P on the circumcirle of $\triangle ABC$, perpendiculars $\overline{PX}, \overline{PY}$ are drawn to lines AC, BC, respectively. P is in the part of arc BC that does not contain point A. If $PX = 45$, $PB = 28$, $PY = 20$, then what is length of PA?

Solution: Note that $\angle PBC = \angle PAC$ implying $PBY \sim PAX$. Thus, $\dfrac{PA}{45} = \dfrac{28}{20}$ so $PA = 63$.

3. What is the sum of the 5-th powers of all the real roots of the equation
$$\log_x(x^3 - 2x^2 - 5x + 14) = 2\ ?$$

Solution: $x > 0$ and $x \neq 1$ because of x is the base of the logarithm function. Then,
$$x^3 - 2x^2 - 5x + 14 = x^2$$

Solutions of AIME Test 4

and
$$x^3 - 2x^2 - 5x + 14 = (x-3)(x+2)(x-2) = 0.$$

Thus, $x \in \{3, 2, -2\}$. Since $x > 0$, we find $x = 3$ or $x = 2$. Therefore, $3^5 + 2^5 = 243 + 32 = 275$.

4. a, b are co-prime positive integers. What is the sum of all the different values

$$\gcd(a+b, a^2 + 17ab + b^2)$$

can take?

Solution: Let $d = \gcd(a+b, a^2 + 17ab + b^2)$. By Euclid's algorithm,
$d = \gcd(a+b, a^2 + 17ab + b^2 - (a+b)^2) = \gcd(a+b, 15ab)$. Well known property that if $\gcd(a,b) = 1$ then $\gcd(a+b, ab) = 1$. Thus, $d = \gcd(a+b, 15ab) = \gcd(a+b, 15)$. Hence $d \in \{1, 3, 5, 15\}$. We can find examples for each value of d. For example, if $a = 1$, $b = 14$ then $d = 15$. So the sum of the d values is $1 + 3 + 5 + 15 = 24$.

5. One of the (x,y) integer pairs satisfying the inequality $|x-y| + |x+y| \leq 20$ is chosen randomly and with equal probability. The probability that $x^2 + y^2 > 35$ is m/n, which m, n co-prime positive integers. What is $m + n$?

Solution: In the analytic plane, graph of the $|x-y| + |x+y| = 20$ is a square. Vertices of the square are $(-10, -10), (-10, 10), (10, -10), (10, 10)$. So, there are $21 \cdot 21 = 441$ lattice points satisfy the inequality $|x-y| + |x+y| \leq 20$. Let's examine $x^2 + y^2 \leq 35$. Then $|x| \leq 5$ and $|y| \leq 5$. We get $11 \cdot 11 = 121$ lattice points. But if $|x| = 5$, then $|y| \neq 4, 5$. If $|x| = 4$, then $|y| \neq 5$. Hence, we find $121 - 2 \cdot 4 - 2 \cdot 2 = 109$ lattice points that satisfy the inequality $x^2 + y^2 \leq 35$. Therefore, the probability is

$$\frac{m}{n} = 1 - \frac{109}{441} = \frac{332}{441}.$$

Solutions of AIME Test 4

$m + n = 332 + 441 = 773.$

6. $ABCD$ is a rectangle and $AD = 10$, $AB = 12$. Points E, F, G, H on the sides AB, BC, CD, DA, respectively such that $AE = DH = 6$, $EH \parallel FG$. The maximum value of the area of the quadrilateral $EFGH$ is m/n, which m, n co-prime positive integers. What is $m + n$?

Solution: Since $EH \parallel FG$, we get $\angle AHE = \angle GFC$ and $\angle AEH = \angle CGF$. Then $AHE \sim CFG$ and $\dfrac{CF}{CG} = \dfrac{AH}{AE} = \dfrac{4}{6}$. Let $CF = 2x$, then $CG = 3x$, $FB = 10 - 2x$, $DG = 12 - 3x$. $Area(ABCD) = 10 \cdot 12 = 120$, $Area(AHE) = \dfrac{1}{2} \cdot 4 \cdot 6 = 12$, $Area(EBF) = \dfrac{1}{2} \cdot 6 \cdot (10 - 2x) = 30 - 6x$, $Area(CGF) = \dfrac{1}{2} \cdot 2x \cdot 3x = 3x$ and $Area(AHE) = \dfrac{1}{2} \cdot 6 \cdot (12 - 3x) = 36 - 9x$. So

$$Area(EFGH) = 120 - (12 + (30 - 6x) + 3x + (36 - 9x)) = -3(x^2 - 5x - 14).$$

The area will be maximum while $x = \dfrac{5}{2}$, because of the area is a parabolic function. Therefore the maxmimum value of $Area(EFGH)$ is $-3(\dfrac{25}{4} - \dfrac{25}{2} - 14) = \dfrac{243}{4}$. Thus, $m + n = 243 + 4 = 247$.

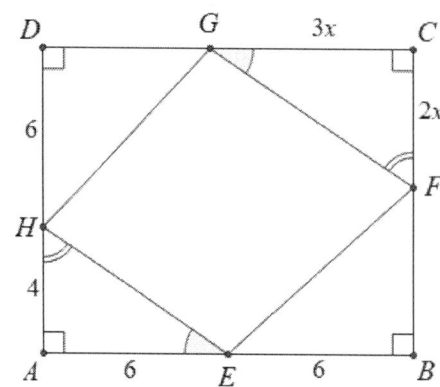

7. In isosceles trapezoid $ABCD$, $AB = 2024$, $CD = 1$ and $AB \parallel CD$. If AC, BC lengths are integer. What is the sum of the different integer values that AC can take?

Solution: Let $AC = BD = x$ and $AD = BC = y$. An isosceles trapezoid is a cyclic quadrilateral. By Ptolemy theorem,

$$x^2 = y^2 + 2024 \cdot 1.$$

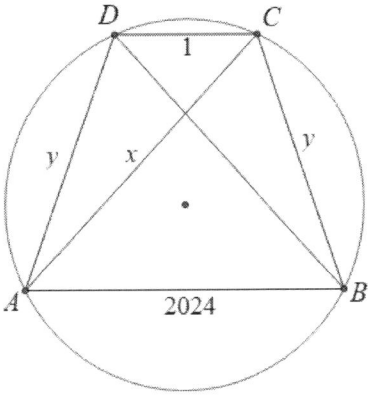

So,

$$x^2 - y^2 = (x-y)(x+y) = 2^3 \cdot 11 \cdot 23.$$

We conclude that $x - y$ and $x + y$ are positive even numbers. Then, all possible cases are

$$\begin{cases} x - y = 2 \\ x + y = 1012 \end{cases} \quad \begin{cases} x - y = 4 \\ x + y = 506 \end{cases} \quad \begin{cases} x - y = 22 \\ x + y = 92 \end{cases} \quad \begin{cases} x - y = 44 \\ x + y = 46 \end{cases}$$

Hence, we get $x \in \{507, 255, 57, 45\}$ and the total is $507 + 255 + 57 + 45 = 864$.

8. We are given n metal weights with positive integer masses (in grams). A brick weighing $k > 0$ grams is glued to the left side of a double-pan balance scale. Our goal is to add some weights to either side of the scale to make it balanced. For example, if we have two weights of 1 gram and 5 grams, and $k = 4$, we can balance the scale by placing the 1 gram weight on the left side and the 5 gram weight on the right side. As it turns out, our n metal weights make it possible to do this for any value of k between 1 and 4000 inclusive. What is the minimum possible value of n?

Solutions of AIME Test 4

Solution: We can use a weight in 3 ways. We either put it on the right side of the scale, we put it on the left side of the scale, or we don't put it on the scale. So we can create 3^n different states for n weights. But one of them is not using any weights at all, which will never balance the scale. This leaves us with $3^n - 1$ states. Also, due to symmetry, at least half of the arrangements of weights do not put more mass on the right side than the left side, which is a requirement for balancing any brick. So we have to divide the remaining states by 2. Thus, using n weights, we can balance at most $\dfrac{3^n - 1}{2}$ different bricks. Thus we get $\dfrac{3^n - 1}{2} \geq 4000$ and $3^n \geq 8001$. Since $3^8 = 6561 < 8001 < 3^9$, we see that $n \geq 9$. Now we prove that 9 weights suffices: the weights we will use are $1, 3, 3^2, \ldots, 3^8$. Now let's complete our solution by proving the following lemma.

Lemma: Given n weights with masses $1, 3, 3^2, \ldots, 3^{n-1}$, we can use them to balance a brick of any integer mass from 1 to $\dfrac{3^n - 1}{2}$ grams glued to the left side of a double pan balance.

Proof: Let's call this proposition $P(n)$, we prove $P(n)$ for all n via mathematical induction.

Base case: If $n = 1$, we have only 1 weight of 1 gram, and only need to prove we can balance the scale when $k = 1$. Clearly we can do this by putting the weight on the right side, so $P(1)$ is true. If we have $n = 2$ weights, 1 gram and 3 grams, we can balance with a brick weighing $1, 3-1, 3$, or $3+1$ gram(s). So we can get values from 1 to $(3^2 - 1)/2 = 4$. $P(2)$ is true.

Inductive step: Now let's assume that $P(m)$ is true. That is, using the m weights $1, 3, 3^2, \ldots, 3^{m-1}$, we can balance any brick weighing $1 \leq k \leq (3^m - 1)/2$ grams glued to a double pan balance. Let's then prove that $P(m+1)$ is true:

- If $1 \leq k \leq (3^m - 1)/2$, then we can balance the scale by the inductive assumption.

- If $(3^m + 1)/2 \leq k \leq 3^m - 1$, start by putting the 3^m weight on the right side of the scale. This will not balance the scale, but instead cause the right side to be heavier than the left side by $3^m - k$ grams. But observe that $1 \leq 3^m - k \leq \frac{3^m - 1}{2}$, so by the inductive assumption we can now balance the scale using some combination of the remaining m weights.

- If $k = 3^m$, we can balance the scale by simply placing the 3^m weight on the right side.

- If $3^m + 1 \leq k \leq \frac{3^{m+1} - 1}{2}$, start by putting the 3^m weight on the right side of the scale. This will not balance the scale, but instead cause the left side to be heaver than the

Solutions of AIME Test 4

Solution: Let $AC = BD = x$ and $AD = BC = y$. An isosceles trapezoid is a cyclic quadrilateral. By Ptolemy theorem,

$$x^2 = y^2 + 2024 \cdot 1.$$

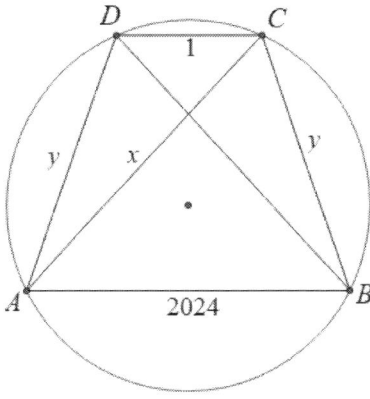

So,

$$x^2 - y^2 = (x-y)(x+y) = 2^3 \cdot 11 \cdot 23.$$

We conclude that $x - y$ and $x + y$ are positive even numbers. Then, all possible cases are

$$\begin{cases} x - y = 2 \\ x + y = 1012 \end{cases} \quad \begin{cases} x - y = 4 \\ x + y = 506 \end{cases} \quad \begin{cases} x - y = 22 \\ x + y = 92 \end{cases} \quad \begin{cases} x - y = 44 \\ x + y = 46 \end{cases}$$

Hence, we get $x \in \{507, 255, 57, 45\}$ and the total is $507 + 255 + 57 + 45 = 864$.

8. We are given n metal weights with positive integer masses (in grams). A brick weighing $k > 0$ grams is glued to the left side of a double-pan balance scale. Our goal is to add some weights to either side of the scale to make it balanced. For example, if we have two weights of 1 gram and 5 grams, and $k = 4$, we can balance the scale by placing the 1 gram weight on the left side and the 5 gram weight on the right side. As it turns out, our n metal weights make it possible to do this for any value of k between 1 and 4000 inclusive. What is the minimum possible value of n?

Solutions of AIME Test 4

Solution: We can use a weight in 3 ways. We either put it on the right side of the scale, we put it on the left side of the scale, or we don't put it on the scale. So we can create 3^n different states for n weights. But one of them is not using any weights at all, which will never balance the scale. This leaves us with $3^n - 1$ states. Also, due to symmetry, at least half of the arrangements of weights do not put more mass on the right side than the left side, which is a requirement for balancing any brick. So we have to divide the remaining states by 2. Thus, using n weights, we can balance at most $\frac{3^n - 1}{2}$ different bricks. Thus we get $\frac{3^n - 1}{2} \geq 4000$ and $3^n \geq 8001$. Since $3^8 = 6561 < 8001 < 3^9$, we see that $n \geq 9$. Now we prove that 9 weights suffices: the weights we will use are $1, 3, 3^2, \ldots, 3^8$. Now let's complete our solution by proving the following lemma.

Lemma: Given n weights with masses $1, 3, 3^2, \ldots, 3^{n-1}$, we can use them to balance a brick of any integer mass from 1 to $\frac{3^n - 1}{2}$ grams glued to the left side of a double pan balance.

Proof: Let's call this proposition $P(n)$, we prove $P(n)$ for all n via mathematical induction.

Base case: If $n = 1$, we have only 1 weight of 1 gram, and only need to prove we can balance the scale when $k = 1$. Clearly we can do this by putting the weight on the right side, so $P(1)$ is true. If we have $n = 2$ weights, 1 gram and 3 grams, we can balance with a brick weighing $1, 3-1, 3,$ or $3+1$ gram(s). So we can get values from 1 to $(3^2 - 1)/2 = 4$. $P(2)$ is true.

Inductive step: Now let's assume that $P(m)$ is true. That is, using the m weights $1, 3, 3^2, \ldots, 3^{m-1}$, we can balance any brick weighing $1 \leq k \leq (3^m - 1)/2$ grams glued to a double pan balance. Let's then prove that $P(m+1)$ is true:

- If $1 \leq k \leq (3^m - 1)/2$, then we can balance the scale by the inductive assumption.

- If $(3^m + 1)/2 \leq k \leq 3^m - 1$, start by putting the 3^m weight on the right side of the scale. This will not balance the scale, but instead cause the right side to be heavier than the left side by $3^m - k$ grams. But observe that $1 \leq 3^m - k \leq \frac{3^m - 1}{2}$, so by the inductive assumption we can now balance the scale using some combination of the remaining m weights.

- If $k = 3^m$, we can balance the scale by simply placing the 3^m weight on the right side.

- If $3^m + 1 \leq k \leq \frac{3^{m+1} - 1}{2}$, start by putting the 3^m weight on the right side of the scale. This will not balance the scale, but instead cause the left side to be heaver than the right side by $k - 3^m$ grams. But observe that $1 \leq k - 3^m \leq \frac{3^m-1}{2}$, so by the inductive assumption we can now balance the scale using some combination of the remaining m weights.

Therefore, we can balance the scale for all $1 \leq k \leq \frac{3^{m+1}-1}{2}$, so $P(m+1)$ is true. This completes the inductive proof, so the previously stated 9 weights do work for all $1 \leq k \leq 4000$, proving the answer to the problem is 9.

9. The tetrahedral sequence of numbers is formed by the number of points: In the first step, there is 1 point. In the $n+1$-th step, a regular tetrahedron with one edge of n units is drawn. Newly added points are on vertices, edges, and surfaces. Among these points, the closest ones are 1 unit apart. The first four tetrahedral numbers are 1, 4, 10, 20, which are shown in the figure below. What term is the number 1873200 in the tetrahedral sequence of numbers?

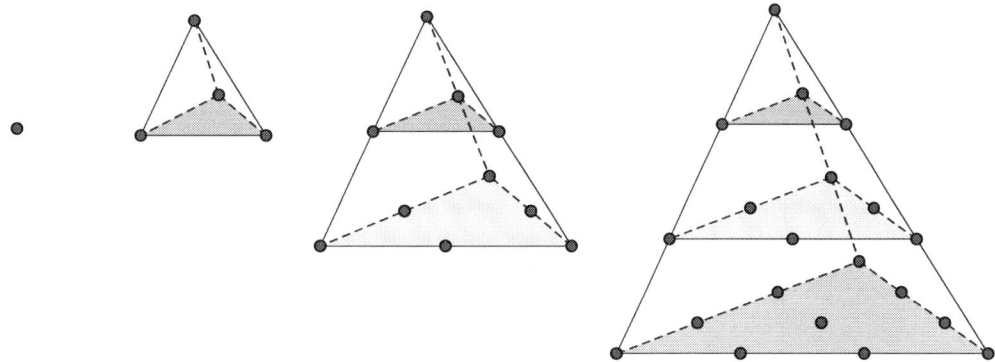

Solution: Let $T(n)$ be tetrahedral sequence. $T(1) = 1$. We see that $T(2) = 1 + (1 + 2)$, $T(3) = 1 + (1 + 2) + (1 + 2 + 3)$, $T(4) = 1 + (1 + 2) + (1 + 2 + 3) + (1 + 2 + 3 + 4)$. Thus,

$$T(n) = 1 + (1+2) + \cdots + (1+2+\cdots+n) = T(n-1) + (1+2+\cdots+n).$$

We can write that

$$T(n) - T(n-1) = \frac{n(n+1)}{2}.$$

From telescoping sum, we find

$$\sum_{k=1}^{n-1}(T(k+1) - T(k)) = \sum_{k=1}^{n-1} \frac{(k+1)(k+2)}{2} = \sum_{k=1}^{n-1} \frac{k^2 + 3k + 2}{2}.$$

Solutions of AIME Test 4

Therefore, $T(n) - T(1) = \dfrac{(n-1)n(2n-1)}{12} + \dfrac{3(n-1)n}{4} + (n-1)$. So, we get that $T(n) = \dfrac{n(n+1)(n+2)}{6}$. If $T(n) = 1873200$, then $n(n+1)(n+2) = 6 \cdot 1873200 = 11239200$. With a little research, we can find that $n = 223$ satisfy the equation.

10. Y is the midpoint of segment XZ. We draw parallel lines x, y, and z through X, Y, and Z respectively so that all 3 lines are perpendicular to XZ. Circle O intersects with x at points A and B, intersects with y at points C and F, and intersects with z at points D and E so that $ABCDEF$ is a convex hexagon. $AB = 2$, $CF = 5$, $DE = 1$, and $XY = \dfrac{a\sqrt{b}}{c}$, where a, b, and c are positive integers with a and c relatively prime and b is not divisible by the square of a prime. Find $a + b + c$.

Solution: Let $XY = x$. We use the pythagorean theorem and Ptolemy's. We compute $BC = \sqrt{x^2 + 1.5^2}$, $BF = \sqrt{x^2 + 3.5^2}$, $EF = \sqrt{x^2 + 4}$, $CE = \sqrt{x^2 + 9}$, and $BE = \sqrt{4x^2 + 1.5^2}$. Now by ptolemy's we get

$$\sqrt{(x^2 + 1.5^2)(x^2 + 4)} + \sqrt{(x^2 + 3.5^2)(x^2 + 9)} = 5\sqrt{4x^2 + 1.5^2}$$

$$\sqrt{(4x^2 + 9)(x^2 + 4)} + \sqrt{(4x^2 + 49)(x^2 + 9)} = 5\sqrt{16x^2 + 9}$$

$$4x^4 + 85x^2 + 441 = (5\sqrt{16x^2 + 9} - \sqrt{(4x^2 + 9)(x^2 + 4)})^2$$
$$= 4h^4 + 425h^2 + 261 - 10\sqrt{(4x^4 + 25x^2 + 36)(16x^2 + 9)}$$

$$100(4x^4 + 25x^2 + 36)(16x^2 + 9) = (340x^2 - 180)^2$$

$$(4x^4 + 25x^2 + 36)(16x^2 + 9) = (34x^2 - 18)^2$$

$$64x^6 + 436x^4 + 801x^2 + 324 = 1156x^4 - 1224x^2 + 324$$

$$64x^4 - 720x^2 + 2025 = (8x^2 - 45)^2 = 0$$

$$x^2 = \dfrac{45}{8}$$

$$x = \dfrac{3\sqrt{10}}{4}$$

Thus the answer is $3 + 10 + 4 = 17$.

11. How many ways are there to color the first 7 positive integers so that each one is colored either red or blue, and there are no 3 integers $1 \leq a < b < c \leq 7$ all the same color forming an arithmetic progression?

    Solution: We bash the solution. WLOG assume 1 is colored red, we will multiply by 2 at the end. Now think about what will happen if 2 is also red:

    $$RRB----$$

    Now there are two cases, if 4 is colored red:

    $$RRBRRBB$$

    or if 4 is colored blue:

    $$RRBBRRB$$
    $$RRBBRBB$$
    $$RRBBRBR$$

    Now we analyze if 2 is colored blue. We first look at the case where 3 is blue:

    $$RBBRRBB$$

    And now we look at if 3 is red:

    $$RBRBBRR$$
    $$RBRRBRB$$
    $$RBRBBRB$$

    In total we find that there are $8 \cdot 2 = 16$ possible ways.

Solutions of AIME Test 4

12. Positive real numbers a, b, c, and d satisfy

$$a + b + c + d = 7$$
$$ab + bc + cd + da = 12$$
$$a^2 + 2b^2 + 3c^2 + 4d^2 = k$$

and the minimum possible value of k is $\dfrac{m}{n}$ where m and n are relatively prime positive integers. Find $m + n$.

Solution: Let $a + c = x$ and $b + d = y$. Then we have that $x + y = 7$ and $xy = 12$, so $\{x, y\} = \{3, 4\}$. Now note that

$$k = a^2 + 2b^2 + 3c^2 + 4d^2 = a^2 + 3(x-a)^2 + 2b^2 + 4(y-b)^2 = 4a^2 - 6ax + 3x^2 + 6b^2 - 8by + 4y^2$$

Which achieves its lowest value when $a = \frac{3x}{4}$ and $b = \frac{2y}{3}$, plugging this in gives

$$k = \frac{3}{4}x^2 + \frac{4}{3}y^2$$

Since we have a choice between $(x,y) = (3,4)$ and $(x,y) = (4,3)$ we plug both of these in and find out the smaller value of k is 24. So our answer is $24 + 1 = 25$.

13. We have $n+1$ different colors of paint. The faces of a regular n-gon pyramid will be painted using these colors. The lateral faces of the pyramid are isosceles triangles. The colorings obtained from one another as a result of the rotation of the object are considered identical. The faces of the pyramid can be painted in $S(n)$ different ways, with each face being a different color.

$$\frac{1}{s(3)} + \frac{1}{S(4)} + \frac{1}{S(5)} + \cdots + \frac{1}{S(2023)} + \frac{1}{2024!} = \frac{m}{n},$$

where m and n relatively co-prime positive integers. What is $m + n$?

Solution: We can paint the base of the pyramid with one of $n+1$ colors. We can paint n lateral surfaces in a circular permutation $(n-1)!$ different ways. So $S(n) = (n+1) \cdot (n-1)!$.

Then,
$$\sum_{n=3}^{2023} \frac{1}{S(n)} = \sum_{n=3}^{2023} \frac{1}{(n+1)\cdot(n-1)!}$$
$$= \sum_{n=3}^{2023} \frac{n}{(n+1)\cdot n\cdot(n-1)!}$$
$$= \sum_{n=3}^{2023} \frac{(n+1)-1}{(n+1)!}$$
$$= \sum_{n=3}^{2023} \left(\frac{1}{n!} - \frac{1}{(n+1)!}\right)$$

and by telescoping sum we get
$$\sum_{n=3}^{2023} \frac{1}{S(n)} = \frac{1}{3!} - \frac{1}{2024!}.$$

Hence,
$$\sum_{n=3}^{2023} \frac{1}{S(n)} + \frac{1}{2024!} = \frac{1}{3!} = \frac{1}{6}$$

and we find $m + n = 1 + 6 = 7$.

14. In triangle ABC with $|BC| = 6$, squares $BADE$, $CBFG$, and $ACHK$ are constructed on the sides of the triangle and extending outward. The points D, E, F, G, H, K are concyclic. Let S be the sum of all distinct possible perimeters of triangle ABC. What is the integer part of S?

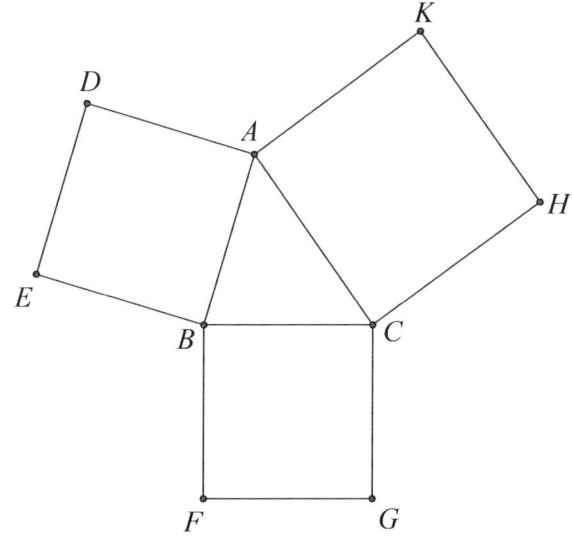

Solution: If the points D, E, F, G, H, K are concyclic, then the center of this circle coincides with the circumcenter of triangle ABC. Let's denote this center as O. If the triangle ABC is equilateral, then the points D, E, F, G, H, K are concyclic. In this case, the perimeter of ABC is $3 \cdot 6 = 18$.

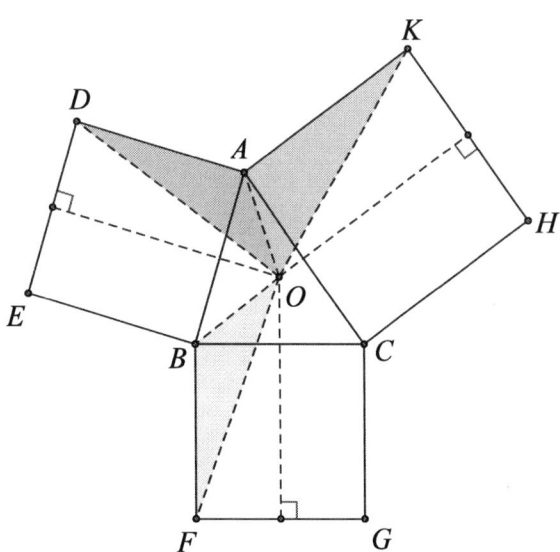

Now, let's consider the cases where at least two sides of triangle ABC are different from each other. Let $|BC| = a$, $|CA| = b$, $|AB| = c$, $\angle A = \alpha$, $\angle B = \beta$, $\angle C = \gamma$, $|OA| = |OB| = |OC| = R$. For this, we can assume $\beta \neq \gamma$. With angle chasing, we find $\angle OBF = 180° - \alpha$, $\angle OAK = 180° - \beta$ and $\angle OAD = 180° - \gamma$. Also, $|BF| = a$, $|AK| = b$, $|AD| = c$. Since point O is the center, we have $|OF| = |OK| = |OD|$. We have the identity $\cos(180° - \theta) = -\cos\theta$. By applying the cosine theorem in triangles OBF, OAK, and OAD, we get $R^2 + a^2 + 2Ra\cos\alpha = R^2 + b^2 + 2Rb\cos\beta = R^2 + c^2 + 2Rc\cos\gamma$. Therefore,

$$a^2 + 2Ra\cos\alpha = b^2 + 2Rb\cos\beta = c^2 + 2Rc\cos\gamma.$$

Solutions of AIME Test 4

By the sine theorem, $a = 2R\sin\alpha$, $b = 2R\sin\beta$, $c = 2R\sin\gamma$. Thus,

$$\sin^2\beta + \sin\beta\cos\beta = \sin^2\gamma + \sin\gamma\cos\gamma$$

$$\sin^2\beta - \sin^2\gamma = \frac{1}{2}(\sin 2\gamma - \sin 2\beta)$$

$$(\sin\beta + \sin\gamma)(\sin\beta - \sin\gamma) = \frac{1}{2}(\sin 2\gamma - \sin 2\beta)$$

$$2\sin\left(\frac{\beta+\gamma}{2}\right)\cos\left(\frac{\beta-\gamma}{2}\right) \cdot 2\cos\left(\frac{\beta+\gamma}{2}\right)\sin\left(\frac{\beta-\gamma}{2}\right) = \cos(\beta+\gamma)\sin(\beta-\gamma)$$

$$\sin(\beta+\gamma)\sin(\beta-\gamma) = -\cos(\beta+\gamma)\sin(\beta-\gamma)$$

$$\tan(\beta+\gamma) = -1$$

and we get that $\beta + \gamma = 135°$, $\alpha = 45°$.

With similar reasoning, from $a^2 + 2Ra\cos\alpha = b^2 + 2Rb\cos\beta$, we find that $\alpha = \gamma$ or $\alpha + \gamma = 135°$. Hence, we conclude that $\alpha = \gamma = 45°, \beta = 90°$ or $\alpha = \beta = 45°, \gamma = 90°$. $|BC| = 6$ and so, the perimeter of ABC is $12 + 6\sqrt{2}$. By symmetry, another solution is $\beta = \gamma = 45°, \alpha = 90°$. For this, the perimeter of ABC is $6 + 6\sqrt{2}$. Therefore,

$$S = 18 + 12 + 6\sqrt{2} + 6 + 6\sqrt{2} = 36 + 12\sqrt{2}$$

and $\lfloor S \rfloor = 36 + 17 = 43$.

15. Let $n \geq 0$ be an integer and let $f(n)$ be the remainder when the smallest prime divisor of $6^{2^n} + 1$ is divided by 32. Find the value of

$$\sum_{n=0}^{100} f(n).$$

Solution: Let's calculate $f(n)$ for the first few values of n.

$n = 0$ yields $6^{2^0} + 1 = 6 + 1 = 7$, and 7 is a prime itself. $f(0) = 7 \pmod{32} = 7$.

$n = 1$ yields $6^{2^1} + 1 = 36 + 1 = 37$, and 37 is a prime itself. $f(0) = 37 \pmod{32} = 5$.

Solutions of AIME Test 4

Let p be the smallest prime divisor of $6^{2^n}+1$. So, we can write $6^{2^n}+1 \equiv 0 \pmod{p}$. $6^{2^n} \equiv -1 \pmod{p}$ and $6^{2^{n+1}} \equiv 1 \pmod{p}$. Hence we conclude that 2^{n+1} is the order of 6 in modulo p. On the other hand, by Fermat's Little Theorem, we have $6^{p-1} \equiv 1 \pmod{p}$. From the properties of order, we obtain that 2^{n+1} divides $p-1$. Thus, for a k positive integer, $p = 2^{n+1}+1$.

$n = 2$ yields $6^{2^2}+1 = 1296+1 = 1297$. If 1297 is a composite number, for a p prime, $p < \sqrt{1297}$ and $p \leq 36$. Also p is in the form $8k+1$. Hence, we have to check $p = 17$ only. But $1297 = 36^2 + 1 \equiv 2^2 + 1 \equiv 5 \not\equiv 0 \pmod{17}$. Therefore 1297 is a prime number. $f(2) = 1297 \pmod{32} = 17$.

$n = 3$ yields $6^{2^3}+1 = 6^8+1$. If 6^8+1 has a p prime divisor, $p = 16k+1$. Let's check $p = 17$. $6^8+1 = 36^4+1 \equiv 2^4+1 \equiv 0 \pmod{17}$. Then, $f(3) = 17 \pmod{32} = 17$.

For $n \geq 4$, if p is a prime divisor of $6^{2^n}+1$, $p = 32k+1$. Thus, $f(n) = 32k+1 \pmod{32} = 1$.

Finally, we get
$$\sum_{n=0}^{100} f(n) = 7 + 5 + 17 + 17 + 97 \cdot 1 = 143.$$

Solutions of AIME Test 5

1. Two of the positive integers from 1 to $2n+1$ (including 1 and $2n+1$) are chosen. The selected numbers can be the same. If the probability is $\dfrac{320}{441}$ of the product of the two chosen numbers is an even number, what is n?

Solution: If the product of the selected numbers is odd, then both of these numbers must be odd. Since there are $n+1$ odd numbers, we can select two odd numbers in $(n+1)^2$ ways. The total number of cases is $(2n+1)^2$. Therefore, the probability that the product is even is $1 - \dfrac{(n+1)^2}{(2n+1)^2} = \dfrac{320}{441}$. From this, we find $\dfrac{(n+1)^2}{(2n+1)^2} = \dfrac{121}{441} = \dfrac{11^2}{21^2}$ and obtain $n = 10$.

2. In the semicircle, $\overline{AB}$ is a diameter with $AB = 8$. Points C and D are on the semicircle such that $BC = CD = 2$ and point C is adjacent to point B. What is $\overline{AD}^2$?

Solution: $ABCD$ is a cyclic quadrilateral. By Thales' theorem, $\angle ACB = \angle ADB = 90°$. Let the intersection of lines AD and BC be point E. Given that $BC = CD = 2$ and $\triangle BDE$ is a right triangle, we have $CE = 2$. In $\triangle ABE$, since $\overline{AC}$ is both an angle bisector and an altitude, $\triangle ABE$ is isosceles. Thus, $AE = AB = 8$. Let $AD = x$. Then, $DE = 8 - x$. In the right triangle $\triangle ABD$, we have $BD^2 = 64 - x^2$. Applying the Pythagorean theorem in the right triangle $\triangle BDE$, we get $EB^2 = BD^2 + DE^2$, which leads to $16 = 64 - x^2 + (8-x)^2$. Solving this equation, we find $x = 7$. Therefore, $\overline{AD}^2 = 7^2 = 49$.

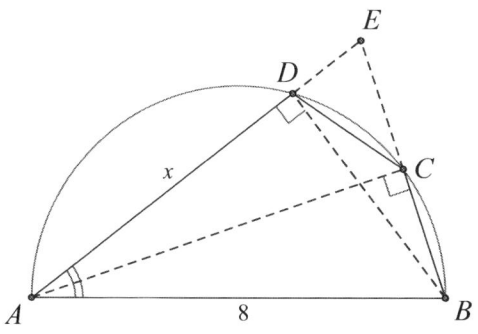

Solutions of AIME Test 5

3. A unit cube is filled with $v < \frac{1}{2}$ cubic units of water. The cube is tilted so that the surface of the water forms an equiangular hexagon $ABCDEF$, such that $AB = CD = EF = 2BC = 2DE = 2FA$. Then, $v = \frac{m}{n}$, where m and n are relatively prime positive integers. Find $m+n$.

Solution: Due to the symmetry of the diagram it is clear that every segment in the hexagon is at a 45° angle with the sides it is adjacent to. Therefore we determine that $AB + BC = \sqrt{2}$ thus we have $AB = \frac{2\sqrt{3}}{3}$ and $BC = \frac{\sqrt{3}}{3}$. Now we realize that the 3D shape we are being asked to find the volume of is a right pyramid with each of the three corners cut off. So we attach right pyramids to segments BC, DE, and FA, creating a right pyramid with side lengths all equal to $\frac{4}{3}$, which has a volume of $(\frac{4}{3})^3 * \frac{1}{6} = \frac{32}{81}$. Each of the three right pyramids we added has side lengths all equal to $\frac{1}{3}$ thus giving them each a volume of $(\frac{1}{3})^3 * \frac{1}{6} = \frac{1}{162}$. Therefore the total volume is $v = \frac{32}{81} - \frac{3}{162} = \frac{61}{162}$, thus making our answer $61 + 162 = 223$.

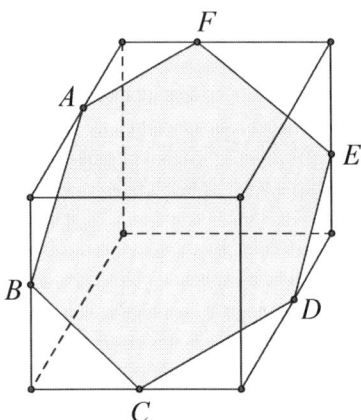

4. Bob makes a bag of red and blue balls as follows:

1. Start with an empty bag.

2. Flip a fair coin to decide between red and blue, then add a ball of the color to the bag.

3. If there are 3 balls in the bag, end the process immediately. Otherwise, flip a fair coin to decide between going back to step 2, or ending this process.

Solutions of AIME Test 5

Bob then hands this bag to Alice. She reaches into the bag and pulls out a ball. It is colored red. She places the ball back into the bag. She then reaches into the bag and pulls out another ball. The probability of this ball being red as well is $\frac{m}{n}$, where m and n are positive integers. Find $m + n$.

Solution: We notice that there are 9 different bags that Bob can create, we list each bag along with the probability of that bag occurring:

- 1 red ball, probability $\frac{1}{4}$.
- 1 blue ball, probability $\frac{1}{4}$.
- 2 red balls, probability $\frac{1}{16}$.
- 1 red and 1 blue ball, probability $\frac{1}{8}$.
- 2 blue balls, probability $\frac{1}{16}$.
- 3 red balls, probability $\frac{1}{32}$.
- 2 red and 1 blue ball, probability $\frac{3}{32}$.
- 1 red and 2 blue balls, probability $\frac{3}{32}$.
- 3 blue balls, probability $\frac{1}{32}$.

Now we list each bag with a nonzero number of red balls in it, along with the probability that Bob creates this bag, AND Alice choose a red ball upon reaching into it:

- 1 red ball, probability $\frac{1}{4}$.
- 2 red balls, probability $\frac{1}{16}$.
- 1 red and 1 blue ball, probability $\frac{1}{16}$.
- 3 red balls, probability $\frac{1}{32}$.
- 2 red and 1 blue ball, probability $\frac{1}{16}$.
- 1 red and 2 blue balls, probability $\frac{1}{32}$.

Notice that all of these probabilities sum to $\frac{1}{2}$. This makes sense since by symmetry there is exactly a $\frac{1}{2}$ chance that Alice chooses a red ball as opposed to a blue ball. Thus, using Bayes' theorem, we can simply multiply each of these probabilities by 2 to obtain the probabilities that Alice has each of these bags, given that she pulled a red ball out at random:

- 1 red ball, probability $\frac{1}{2}$.
- 2 red balls, probability $\frac{1}{8}$.
- 1 red and 1 blue ball, probability $\frac{1}{8}$.

Solutions of AIME Test 5

- 3 red balls, probability $\frac{1}{16}$.
- 2 red and 1 blue ball, probability $\frac{1}{8}$.
- 1 red and 2 blue balls, probability $\frac{1}{16}$.

Thus, we can now simply multiply each of these probabilities by the probability that Alice will pull a red ball out again, and sum everything to get our final answer, A:

$$A = \frac{1}{2} + \frac{1}{8} + \frac{1}{16} + \frac{1}{16} + \frac{1}{12} + \frac{1}{48} = \frac{41}{48}$$

Thus our final answer is $41 + 48 = 89$.

5. A 6×5 table containing the natural numbers from 1 to 30 has been provided. One number is selected from each column and these five numbers are multiplied together. For example, $1 \cdot 2 \cdot 8 \cdot 9 \cdot 30 = 4320$. The sum of all these products is denoted as S. What is the remainder when S is divided by 1000?

1	2	3	4	5
6	7	8	9	10
11	12	13	14	15
16	17	18	19	20
21	22	23	24	25
26	27	28	29	30

Solution: An important idea is

$$S = (1+6+11+16+21+26)(2+7+12+17+22+27)\cdots(5+10+15+20+25+30).$$

Also,

$$1+6+11+16+21+26 = \frac{1+26}{2} \cdot 6 = 27 \cdot 3,$$
$$2+7+12+17+22+27 = \frac{2+27}{2} \cdot 6 = 29 \cdot 3,$$

$$3+8+13+18+23+28 = \frac{3+28}{2} \cdot 6 = 31 \cdot 3,$$
$$4+9+14+19+24+29 = \frac{4+29}{2} \cdot 6 = 33 \cdot 3,$$
$$5+10+15+20+25+30 = \frac{5+30}{2} \cdot 6 = 35 \cdot 3$$

Therefore $S = 27 \cdot 3 \cdot 29 \cdot 3 \cdot 31 \cdot 3 \cdot 33 \cdot 3 \cdot 35 \cdot 3$. Also, $29 \cdot 35 = 32^2 - 3^2 = 1015 \cdot 1033$, $31 \cdot 33 = 32^2 - 1^2 = 1023 \cdot 1025$. Hence, $S = 3^8 \cdot 1015 \cdot 1033 \cdot 1023 \cdot 1025 \equiv 3^8 \cdot 15 \cdot 33 \cdot 23 \cdot 25$ (mod 1000). Hence, we get $S \equiv 545$ (mod 1000).

6. Let ABC be a triangle, and let D be the midpoint of $\overline{BC}$. The incircle ω of $\triangle ABC$ intersects $\overline{AD}$ at two distinct points, which trisect the median $\overline{AD}$. Circle ω is tangent to $\overline{BC}$ at point E. Given that $ED = 8$, and D is between points E and C. If the area of ω is $\frac{a\pi}{b}$, where a and b relatively co-prime positive integers, what is $a+b$?

Solution: Let $BC = a$, $CA = b$, and $AB = c$. Also let the incircle be tangent to CA at F and AB at G. Also let AD intersect the incircle at points X and Y such that X is between A and Y. Now note that $AX * AY = YD * XD$ thus $AG^2 = ED^2$, so $AG = 8$. Then since $BG = BE$, we know $BA = BD$ implying $a = 2c$. Let $BG = BE = x$. Then $AB = 8 + x$, $BC = 16 + 2x$, and $CA = 24 + x$. Also since $AX * AY = AG^2 = 64$ and $AY = 2AX$, we determine that $AX = 4\sqrt{2}$ and thus $AD = 12\sqrt{2}$. We can also drop the perpendiculars from B to AD and C to AD, let the feet of these altitudes be J and K respectively. It is easy to see that BDJ and CDK are congruent triangles, thus since J is the midpoint of AD we get $AK = 18\sqrt{2}$. Thus we see that $AC^2 - 648 = DC^2 - 72$ so $(24+x)^2 - (8+x)^2 = 576$ thus $32 + 2x = 36$ so $x = 2$. Thus the side lengths of the triangle are 10, 20, and 26, so using Heron's formula we determine the triangle's area to be $\sqrt{28(18)(8)(2)} = 28r$ thus $r = \frac{6\sqrt{14}}{7}$ so we determine $r^2 = \frac{72}{49}$ thus giving us the answer $72 + 49 = 121$.

Solutions of AIME Test 5

7. Given that $(x^{2022} + x^3 - 1)^{2023} = a_0 + a_1 x + a_2 x^2 + a_3 x^3 + \cdots + a_{2022 \cdot 2023} x^{2022 \cdot 2023}$. Find the sum

$$S = a_0 + a_3 + a_6 + \cdots + a_{2022 \cdot 2023}.$$

Solution: $x^2 + x + 1 = 0$ equation has complex roots x_1 and x_2. $x_1^3 = x_2^3 = 1$ because $x^3 - 1 = (x-1)(x^2 + x + 1) = 0$ and $x^3 = 1$. Also, $x_2 = x_1^2$ and $x_1 = x_2^2$. Easily, we can see that $x_1^{3k+1} + x_1^{3k-1} + 1 = x_2^{3k+1} + x_2^{3k-1} + 1 = 0$, where k in an integer. If we substitute 1, x_1, and x_2 for x in the given equation respectively, we obtain the equations:

$$(1 + 1 - 1)^{2023} = a_0 + a_1 + a_2 + a_3 + \cdots + a_{2022 \cdot 2023}$$

$$(x_1^{2022} + x_1^3 - 1)^{2023} = a_0 + a_1 x_1 + a_2 x_1^2 + a_3 x_1^3 + \cdots + a_{2022 \cdot 2023} x_1^{2022 \cdot 2023}$$

$$(x_2^{2022} + x_2^3 - 1)^{2023} = a_0 + a_1 x_2 + a_2 x_2^2 + a_3 x_2^3 + \cdots + a_{2022 \cdot 2023} x_2^{2022 \cdot 2023}$$

From the sum of these equations, we obtain

$$3 = 3 \left(a_0 + a_3 + a_6 + \cdots + a_{2022 \cdot 2023} \right).$$

Thus, we find $S = 1$.

8. Given that m and n are positive real numbers, there is no real number pair (x, y) that satisfies the system of equations:

$$x^3 + y^3 = 16$$

$$mx + y = n$$

What is the smallest positive integer value that $m^2 + n^2$ can take?

Solution: Let's substitute $y = -mx + n$ in the equation $x^3 + y^3 = 16$. We get $x^3 + (-mx + n)^3 = 16$. So, $(1 - m^3)x^3 - 3m^2 n x^2 + 3mn^2 x - n^3 - 16 = 0$. We know that a cubic equation has a real root. Then, the last equation can not be cubic. Hence $1 - m^3 = 0$ and we find $m = 1$. Thus we have the quadratic equation

$$3nx^2 - 3n^2x + n^3 + 16 = 0.$$

x is not a real number, so the discriminant have to be negative. $\Delta = 9n^4 - 4 \cdot 3n \cdot (n^3 - 16) < 0$. Since $n > 0$, we can write $9n^3 - 4 \cdot 3 \cdot (n^3 - 16) < 0$. Therefore $n^3 > 64$ and $n > 4$. $m^2 + n^2 > (-1)^2 + 4^2 = 17$. Thus, the minimum integer value of $m^2 + n^2$ is 18. We can take $n = \sqrt{17}$.

9. A student plays a game as follows. They first choose a real number p between 0 and 1, inclusive. They then receive a weighted coin that lands heads with probability p and tails with probability $1 - p$. They flip it 7 times, and they win if it comes up heads exactly 3 times. The value of p the student should choose to maximize their odds of winning is $\frac{m}{n}$, where m and n are relatively prime positive integers. Find $m + n$.

Solution: Notice there are $\binom{7}{3} = 35$ sequences of coin flips with 3 heads and 4 tails in some order, and each of them has probability $p^3(1-p)^4$ of happening, so the total probability of winning is $35p^3(1-p)^4$. Thus we wish to maximize $p^3(1-p)^4$. Now the critical step: Multiply this expression by $\frac{64}{27}$ to get $\left(\frac{4p}{3}\right)^3 (1-p)^4$. Then by AM-GM inequality we have $\sqrt[7]{\left(\frac{4p}{3}\right)^3(1-p)^4} \leq \frac{\frac{4p}{3}+\frac{4p}{3}+\frac{4p}{3}+(1-p)+(1-p)+(1-p)+(1-p)}{7} = \frac{4}{7}$ with equality specifically when $\frac{4p}{3} = 1 - p$ thus $p = \frac{3}{7}$. So our answer is $3 + 7 = 10$.

10. We have that a and b are positive real numbers such that $a + b = 2408$ and

$$\log_a(\log_b(a)) = \log_b(\log_a(\sqrt[1024]{b}))$$

Find $b - 300a$.

Solutions of AIME Test 5

Solution: Let $b = a^x$. Then the equation simplifies as follows:

$$\log_a\left(\frac{1}{x}\right) = \log_b\left(\frac{x}{1024}\right)$$

$$\log_a\left(\frac{1}{x^x}\right) = \log_b\left(\left(\frac{x}{1024}\right)^x\right) = \log_a\left(\frac{x}{1024}\right)$$

$$x^{(x+1)} = 1024 = 4^5$$

Thus $x = 4$. Then $a + a^4 = 2408 = 7^4 + 7$ thus $a = 7$, $b = 2401$, and $b - 300a = 301$.

11. Let $ABCD$ be a tetrahedron, and M be the midpoint of AB. We have $CD = 14$, $CA = 13$, $AD = 15$, $CM = 5$, and $MD = \sqrt{109}$. Find the largest possible integer value that the perimeter of triangle CBD can be.

Solution: Drop the altitudes from A to CD, M to CD, and B to CD. Let the feet of these altitudes be X, Y, and Z respectively. Let the length of these altitudes be h_1, h_2, and h_3 respectively.

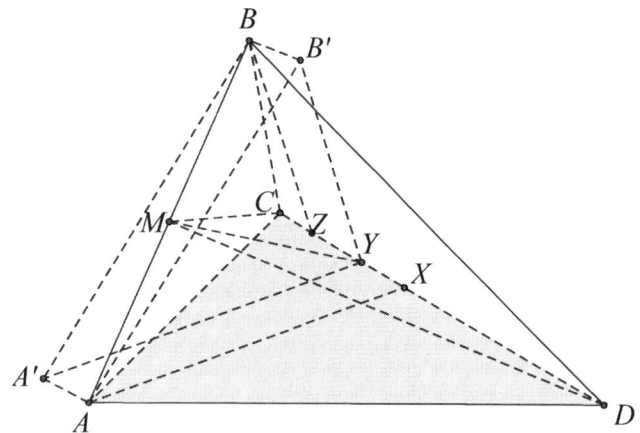

Then, construct points A' and B' such that $AA'YX$ and $BB'YZ$ are rectangles. Since $A'Y \perp CD$, $MY \perp CD$, and $B'Y \perp CD$, we have that A', M, B', and Y lie in the plane through Y perpendicular to line CD. Furthermore notice that $BB' \| CD \| AA'$ so $AA'BB'$ is a trapezoid, $AA'MBB'$ are co-planar, and we know $BB' \perp B'M$ and $AA' \perp A'M$. Finally we know that $\angle A'AM = \angle B'BM$ by the parallel lines and $AM = MB$, therefore triangles

Solutions of AIME Test 5

$A'AM$ and $B'BM$ are congruent by AAS, thus $A'A = BB'$ and $A'AB'B$ is a parallelogram implying that M is the midpoint of $A'B'$. Now, we can find the area of triangle ACD with side lengths 13, 14, and 15 using Heron's formula to be 84. From the area relation $\frac{h_1 \cdot 14}{2} = 84$, we obtain $h_1 = 12$. Similarly, in triangle MCD, we find $h_2 = 3$. Using the Pythagorean theorem we now see that $CX = 5$ and $CY = 4$, and since $AM = MB$, $XY = YZ$, so $CZ = 3$. Thus, the length of h_3 uniquely determines triangle CBD, and it is easy to see that since the perimeter of CBD is $14 + \sqrt{3^2 + h_3^2} + \sqrt{11^2 + h_3^2}$, in order to maximize the perimeter we want to maximize h_3. This is where A' and B' become very useful, as $A'Y = 12$, $MY = 3$, YM is a median of triangle $A'B'Y$, and we are trying to maximize the length of $B'Y$. Notice that by triangle inequality, $MB' = A'M < A'Y + YM = 15$, thus $h_3 = YB' < YM + MB' < 18$. If A', Y, and M are collinear with Y in between A' and B', we can see that $YB' = 18$. Thus h_3 is at most 18, although if $h_3 = 18$ then the tetrahedron becomes degenerate. However, whether you consider a degenerate tetrahedron to still be a tetrahedron does not affect the answer to the problem, as when $h_3 = 18$ the perimeter of CBD is not an integer. Note $14 + \sqrt{3^2 + 18^2} + \sqrt{11^2 + 18^2} = 14 + \sqrt{333} + \sqrt{445}$ can be approximated to be about $14 + 18.2 + 21.1$ which is a little over 53, thus the maximum integer value of the perimeter must be 53.

12. Let the Fibonacci numbers be defined $F_1 = 1$, $F_2 = 1$, and $F_{n+2} = F_{n+1} + F_n$ for all $n \geq 1$. If
$$\sum_{n=5}^{\infty} \frac{1 + (-1)^n F_n F_{n+2}}{F_n F_{n+1} F_{n+2}} = \frac{m}{n}$$
Where m and n are relatively prime integers, find $-mn$.

Solution: First we show that $F_{n+3}F_n - F_{n+2}F_{n+1} = (-1)^{n+1}$ via induction: Note that $F_3 F_0 - F_2 F_1 = 0 - 1 = -1$, and that for all $n \geq 1$, $F_{n+3}F_n - F_{n+2}F_{n+1} = F_{n+2}F_n + F_{n+1}F_n - F_{n+2}F_n - F_{n+2}F_{n-1} = F_{n+1}F_n - F_{n+2}F_{n-1} = -1 * (-1)^n = (-1)^{n+1}$. Thus we have that

$$\sum_{n=5}^{\infty} \frac{1 + (-1)^n F_n F_{n+2}}{F_n F_{n+1} F_{n+2}} = \sum_{n=5}^{\infty} (-1)^n \frac{F_{n+2}F_{n+1} - F_{n+3}F_n + F_n F_{n+2}}{F_n F_{n+1} F_{n+2}} = \sum_{n=5}^{\infty} (-1)^n \frac{F_{n+2}F_{n+1} - F_{n+1}F_n}{F_n F_{n+1} F_{n+2}}$$

93

Solutions of AIME Test 5

$$= \sum_{n=5}^{\infty}(-1)^n \frac{F_{n+2} - F_n}{F_n F_{n+2}} = \sum_{n=5}^{\infty}(-1)^n \left(\frac{1}{F_n} - \frac{1}{F_{n+2}}\right)$$

which is a telescoping sum which simply equals $-\frac{1}{F_5} + \frac{1}{F_6} = \frac{-3}{40}$ making our answer 120.

13. In Triangle ABC with $BC = 5$, $CA = 27$, and $\angle ABC = 90°$, let E be a point on segment CA. Let ω be the circle passing through B and tangent to CA at E. ω intersects segment BC at a point D not equal to B, and we have that $BD = DE = \frac{m}{n}$, where m and n are relatively prime positive integers. Find $m + n$.

Solution: Let $BD = x$. Draw segments BE and DE. Notice that $DE = x$ and $DC = 5 - x$. Also, because the circumcircle of BDE is tangent to EC, we have that $\angle DBE = \angle DEC$ and thus triangles DEC and EBC are similar.

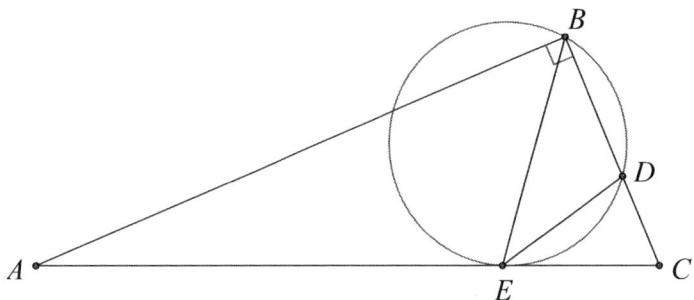

Thus we have $\frac{DC}{EC} = \frac{EC}{BC}$ thus $EC = \sqrt{BC \cdot DC} = \sqrt{25 - 5x}$. Finally, using the law of cosines on triangle DCE, since $\cos \angle DCE = \frac{5}{27}$,

$$x^2 = (5-x)^2 + 25 - 5x - 2(5-x)^{\frac{3}{2}} \cdot \frac{5\sqrt{5}}{27}$$

$$10 - 3x = (5-x)^{\frac{3}{2}} \cdot \frac{2\sqrt{5}}{27}$$

$$27^2(9x^2 - 60x + 100) = 20(5-x)^3$$

$$20x^3 + 6261x^2 - 42240x + 70400 = 0$$

Since we know the root is rational, we can use the rational root theorem to find the root. However, the issue here is that $70400 = 2^8 \cdot 5^2 \cdot 11$ has a lot of factors, so it would be unwise to simply try every single one. Instead, we first bound the value of x. First since D

Solutions of AIME Test 5

lies on segment BC, we know $0 < x < 5$. This tells us that $5 - x$ is positive, thus since $10 - 3x = (5-x)^{\frac{3}{2}} \cdot \frac{2\sqrt{5}}{27}$, we know that $x < \frac{10}{3}$. Also we can calculate $f(3) > 0$ and $f(\frac{10}{3}) < 0$, thus we see that the root lies between 3 and $\frac{10}{3}$. This significantly reduces our search space as most of the rational numbers with numerator dividing 70400 and denominator dividing 20 are very large. The only two rational numbers that satisfy the divisibility and bound properties needed are $\frac{16}{5}$ and $\frac{25}{8}$. Testing both out we see $\frac{16}{5}$, and plugging into our original equation for x we see it is not an extraneous solution. So our answer is $16 + 5 = 21$.

Note: We can also divide out $(5x - 16)$ from the cubic and solve the remaining quadratic, but the roots of the quadratic equation in the cubic $(5x - 16)(4x^2 + 1265x - 4400) = 0$ are irrational.

14. Non-constant polynomials $P(x)$, $Q(x)$, and $R(x)$ with integer coefficients satisfy the following conditions:

    - All 3 polynomials have the same degree.
    - For all real numbers x, $(P(x))^2 + (Q(x))^2 = (R(x))^2$.
    - For all real numbers x, $P(Q(x)) - P(0) + Q(P(x)) - Q(0) + 4x = R(R(x)) - R(0) + 6$.

Find the sum of the three smallest values of $|R(0)|$.

Solution: First, assume the polynomials have degree greater than 1. We show there are no solutions. Let the leading coefficient of $P(x)$ be p, of $Q(x)$ be q, and of $R(x)$ be r. By definition, none of these 3 numbers can equal 0. Furthermore let the degree of all three polynomials be $d > 1$. Then we have from the first equation that $p^2 + q^2 = r^2$ and from the second equation that $pq^d + p^d q = r^{d+1}$. Since both of these equations are homogeneous, we can WLOG assume $\gcd(p, q, r) = 1$. Now, say there is some prime p_0 that divides p. From the second equation, since $p_0 | pq^d + p^d q$ it is also the case that $p_0 | r^{d+1}$ so $p_0 | r$. But then from the first equation, p_0 also divides q, contradicting our assumption that all three numbers are relatively prime. Thus $p = \pm 1$ and by symmetry $q = \pm 1$ as well. Thus $r = \sqrt{2}$ which isn't

Solutions of AIME Test 5

an integer, contradiction, so no solutions exist.

Now let $P(x) = ax + b$, $P(x) = cx + d$, and $R(x) = ex + f$. The first equation gives us $a^2 + c^2 = e^2$, $ab + cd = ef$, and $b^2 + d^2 = f^2$. The second equation gives us $2ac + 4 = e^2$ and $ad + bc = ef + 6$. Note this means $a^2 + c^2 - 2ac - 4 = 0$ so $a - c = \pm 2$. Since everything in the problem is symmetric in P and Q, WLOG let $c = a + 2$. Also, $ad + bc - ab - cd = 6$ but this can be factored as $(a - c)(d - b) = 6$. Therefore $b = d + 3$. Now note that $(a^2 + c^2)(b^2 + d^2) = e^2 f^2 = (ab + cd)^2$, but by expanding and canceling terms we get $b^2 c^2 + a^2 d^2 - 2abcd = 0$ which factors as $(bc - ad)^2 = 0$ so $bc = ad$. Plugging in $d = b - 3$ and $c = a + 2$ we get $2b = -3a$. Let $a = 2k$ so $b = -3k$. Then $c = 2k + 2$ and $d = -3k - 3$. Finally since $b^2 + d^2 = f^2$ we see that $f = \pm 3\sqrt{k^2 + (k+1)^2}$ for some integer k. By trying nonnegative values of k we see that $k = 0$, $k = 3$, and $k = 20$ give us the smallest integer values of $|f|$ which are 3, 15, and 87, giving us a final answer of 105.

15. Starting from vertex A of a regular n-gon with a side length of 1 unit, moving 1 unit along the edges in the positive direction leads to point B, moving 2 units leads to point C, moving 4 units leads to point E, and moving 5 units leads to point F. (It is possible to return to points A, B, or C as a result of the movement.) Given that the areas of triangles ABF and ACE are equal, what is the sum of the possible values of n?

Solution: In the case $n = 3$, if we draw the equilateral triangle ABC, both ABF and ACE will also be equilateral triangles, satisfying the condition. It's straightforward to see that no such area equality occurs for $n = 4$ or $n = 5$.

Let $n \geq 6$. Now, the situation of 'wrapping around the polygon' will no longer occur. Let's label the regular polygon as $ABCDEF\ldots$. Drawing the circumscribed circle with radius R, let α represent the central angle subtended by each side. $0 < \alpha \leq 30°$. Naturally, one exterior angle measure of the regular polygon is $\theta = 2\alpha$. We have $\angle AFB = \alpha$ and $\angle FAB = 4\alpha$.

Using the formula for the area of a triangle as one-fourth the product of its sides divided by the radius of its circumscribed circle, along with the sine theorem, we can write $\text{Area}(ABF) =$

Solutions of AIME Test 5

$2R^2 \sin(\alpha) \sin(4\alpha) \sin(5\alpha)$. Similarly, $Area(ACE) = 2R^2 \sin(2\alpha) \sin(2\alpha) \sin(4\alpha)$.

Given that $Area(ABF) = Area(ACE)$, we get

$$\sin(\alpha) \sin(4\alpha) \sin(5\alpha) = \sin(2\alpha) \sin(2\alpha) \sin(4\alpha)$$

Since $0 < \alpha \leq 30°$, $\sin(4\alpha) > 0$. Simplifying by $\sin(4\alpha)$, we have

$$\sin(\alpha) \sin(5\alpha) = \sin(2\alpha) \sin(2\alpha)$$

Using the identity $\sin(x) \sin(y) = \frac{1}{2}(\cos(x-y) - \cos(x+y))$,

$$\cos(4\alpha) - \cos(6\alpha) = \cos(0) - \cos(4\alpha)$$

$$\cos(2\theta) - \cos(3\theta) = 1 - \cos(2\theta)$$

Using the trigonometric identities $\cos(2\theta) = 2\cos^2(\theta) - 1$ and $\cos(3\theta) = 4\cos^3(\theta) - 3\cos(\theta)$, and letting $x = \cos(\theta)$, we arrive at

$$4x^3 - 4x^2 - 3x + 3 = 0$$

$$(x-1)(4x^2 - 3) = 0$$

Solving this equation in the interval $0 < x < 1$, we find $x = \cos(\theta) = \frac{\sqrt{3}}{2}$, and hence $\theta = 30°$. The number of sides of the polygon is $n = \frac{360}{30} = 12$.

Therefore, the sum of all possible values for n is $3 + 12 = 15$.

Made in the USA
Las Vegas, NV
14 January 2025

16323610R00061